Belmont Revisited

Belmont Revisited

Ethical Principles for Research with Human Subjects

JAMES F. CHILDRESS, ERIC M. MESLIN,
AND HAROLD T. SHAPIRO, EDITORS

GEORGETOWN UNIVERSITY PRESS
Washington, DC

Georgetown University Press, Washington, D.C.
© 2005 by Georgetown University Press. All rights
reserved.

Printed in the United States of America

10 9 8 7 6 5 4 3 2 1   2005

Library of Congress Cataloging-in-Publication Data

Belmont revisited : ethical principles for research with
human subjects / James F. Childress, Eric M. Meslin &
Harold Shapiro, editors.
    p.   cm.
  Includes bibliographical references and index.
    ISBN 1-58901-062-0 (pkb. : alk. paper)
1. Human experimentation in medicine — Moral and
ethical aspects.   2. Medical ethics.   3. Medicine —
Research — Moral and ethical aspects.   I. Childress,
James F.   II. Meslin, Eric Mark.   III. Shapiro, Harold.
    R853.H8B45   2005
    174.2 — dc22

Contents

Introduction

Which principles should govern biomedical research and practice? Over twenty-five years ago the National Commission for the Protection of Human Subjects of Biomedical and Behavioral Research addressed this question with reference to biomedical and behavioral research, as required by its legislative mandate: "to identify the basic ethical principles that should underlie the conduct of biomedical and behavioral research involving human subjects and to develop guidelines which should be followed to assure that such research is conducted in accordance with those principles." The result was the *Belmont Report: Ethical Principles and Guidelines for the Protection of Human Subjects of Research*, published in the *Federal Register* on April 18, 1979.[1] Although the *Belmont Report* made several important conceptual points about research ethics, it is best known for its presentation and discussion of three principles — respect for persons, beneficence, and justice — and the way in which these principles inform specific guidelines for research involving human subjects: for informed consent, risk-benefit assessment, and selection of subjects. These principles and guidelines remain an important, though incomplete, ethical framework for research involving human subjects, especially in the United States. In addition, they have been extended well beyond their initial context into medical practice and health care.

The National Commission, which existed from 1974 to 1978, was succeeded by a series of other national bodies in public bioethics. One was the National Bioethics Advisory Commission (NBAC), 1996–2001, which was appointed by President Clinton. As chair (Shapiro), commissioner (Childress), and executive director (Meslin) of the NBAC, we and other commissioners and staff spent some time discussing whether it would be useful to revisit the *Belmont* principles and perhaps reinterpret, modify, or augment them for the twenty-first century. Indeed, at NBAC's very first meeting, one commissioner, Ezekiel Emanuel, recommended that a fourth principle — respect for communities — might usefully be added to the *Belmont* triplet to accommodate broader social considerations. However, in view of NBAC's other urgent tasks, which were greatly enlarged as a result of breakthroughs in cloning and human embryonic stem cell research, as well as the difficulty of preparing a document that could supplant the *Belmont Report*, another alternative became more attractive. The commissioners were invited to participate with members and staff of previous U.S. bioethics commissions and other experts in research ethics in a conference in the spring of 1999, cosponsored by Princeton University and the University of Virginia's Institute for Practical Ethics. This conference, which coincided with the twentieth anniversary of the publication of the *Belmont Report* in the *Federal Register*, provided a welcome venue to revisit the *Belmont* principles, to trace their interpretation by other national bodies in public bioethics, to consider their impact outside research involving human subjects, and to assess their adequacy, in both content and application, for research involving human subjects and other domains. The following essays, which were initially presented as outlines or drafts at the "*Belmont* Revisited" conference, held at the University of Virginia, have been further developed for publication in this volume.

The first two chapters, which focus on the origins of the *Belmont Report* and its three principles, were written by two ethicists who played an important role in their development. Albert R. Jonsen, a bioethicist and member of the National Commission, as well as professor emeritus of the University of Washington School of Medicine, recalls how the principles emerged in the commission's deliberations, notes various influences on the selection and formulation of these principles, and reflects on the future of what, given its purpose at the time, he views as "a most satisfactory document," despite several limitations. In line with part of the congressional mandate that established the National Commission, the purpose of this report was "to provide to the research community . . . a clear, concise statement of the principles that, if observed in particular practices, would render their enterprise ethical."

Philosopher Tom L. Beauchamp (Kennedy Institute of Ethics and Department of Philosophy at Georgetown University) then offers an account of the development of the *Belmont Report* from his perspective as a staff member of the National Commission, beginning in late 1976 after the commissioners had already settled on the three basic principles. Beauchamp also discusses its relation to *Principles of Biomedical Ethics*, which he was writing with James F. Childress at the same time. These two works — *Belmont* and *Principles of Biomedical Ethics* — he stresses, represent "substantially different moral visions." Jonsen's and Beauchamp's perspectives, despite some differences, together offer rich insight into the emergence and evolution of the *Belmont* principles in the larger context of the new and developing field of bioethics.

Jonsen indicates that he and his colleagues had imagined that an Ethics Advisory Board (EAB) would be established to provide interpretations and applications of the *Belmont* principles over time. For various reasons, the EAB was short-lived (1978–80). Succeeding both the National Commission and the EAB was a variety of national bodies that sometimes incorporated the *Belmont* principles, sometimes modified them, sometimes supplemented them, and sometimes used different terms and concepts. This volume devotes chapters to three of these bodies in public bioethics, written by their members and/or staff.

The first of these bodies discussed is the President's Commission for the Study of Ethical Problems in Medicine and Biomedical and Behavioral Research, which existed January 1980–March 1983. The chapter by Alexander M. Capron, a lawyer and bioethicist now at the University of Southern California, who was executive director of the President's Commission, reflects on "the curious relationship" between the *Belmont Report* and the work of the President's Commission. According to Capron, the *Belmont Report* "played virtually no part in the deliberations or conclusions of the President's Commission." There are several possible explanations for this. One focuses on the different and broader scope of the President's Commission's responsibilities — "to examine the ethical, social, and legal issues in health care rather than in research." Capron, however, finds a more plausible explanation in the judgment by the "staff ethicists" that the *Belmont* principles were "overly broad, vaguely defined, and unranked" and that they maintained a preference for "general normative theory" rather than bioethical theory, insofar as they drew on theory at all. A final explanation can be found in the inductive approach and contextual analysis used by the President's Commission (as well as by the earlier National Commission) as it considered philosophical and ethical issues embedded in the topics being addressed. Nevertheless, as Capron notes, the *Belmont* principles or similar ones are used in some of the President's Commission's reports. One example is the appeal to principles of self-determination, well-being, and equity in the President's Commission's report on *Deciding to Forego Life-Sustaining Treatment* (March 1983).

Next, the relationship between the *Belmont* principles and the work of the Advisory Committee on Human Radiation Experiments (ACHRE), from 1994 to 1995, is subjected to examination by the ACHRE chair, Ruth R. Faden (director, The Phoebe Berman Bioethics Institute, the Johns Hopkins University) and two members of its staff, Anna C. Mastroianni (University of Washington School of Law), and Jeffrey P. Kahn (director, Center for Bioethics, University of Minnesota). ACHRE was established by President Clinton to investigate human radiation experiments (the majority of which were carried out under the auspices of the U.S. military) and deliberate releases of radiation into the environment, from 1944 to 1974. This historical research was augmented by further scholarship to identify the ethical and scientific standards for evaluating these actions and practices as well as to propose measures to prevent their recurrence. The authors note that "the influence of the *Belmont Report* is evident in aspects of [ACHRE's] historical evaluation, and is prominent in [its] evaluation of the current state of protections for human subjects." In addition, ACHRE identified six basic ethical principles: "One ought not to treat people as mere means to the ends of others"; "One ought not to deceive others"; "One ought not to inflict harm or risk of harm"; "One ought to promote welfare and prevent harm"; "One ought to treat people fairly and with equal respect"; and "One ought to respect the self-determination of others." Although these principles could be grouped under the *Belmont* principles, ACHRE did not choose to do so (and in this way may have been similar to the President's Commission), preferring instead to have several principles that were all (arguably) applicable during the period of the research in question. Although ACHRE developed "a healthy skepticism about regulations and rules," it continued to recognize "the importance of moral principles as public *expressions of moral commitment*, expressions that bind the nation and the professions in a common moral vision for the enterprise of human research." This view fits well with ACHRE's attention to and emphasis on two other themes — trust and openness — that go well beyond the *Belmont Report* and that grew out of ACHRE's own historical and contemporary analyses.

Another body in public bioethics, which in part resulted from ACHRE's recommendations, was NBAC, 1996–2001. Its chair, Shapiro, and its executive director, Meslin, discuss NBAC's various reports in light of the *Belmont* principles, noting that these principles played only a modest role in some reports, such as human cloning, a mixed role in others, such as human embryonic stem cell research, but a very prominent and influential role in others, particularly the reports that addressed research involving human participants. NBAC did not simply apply the *Belmont* principles but rather reinterpreted and expanded them as appropriate. This was particularly evident in NBAC's argument, based on an expanded conception of justice in research, that U.S.-conducted or -sponsored research in other countries should be responsive to the health needs of those countries.

The President's Commission, ACHRE, and NBAC are all examples of public bodies grappling with bioethics. Bioethical principles, identical or similar to the *Belmont* principles, were becoming increasingly prominent and prevalent in medicine and health care. Eric Cassell — another member of NBAC, a physician, and a professor emeritus at the Cornell University School of Medicine — presents two anecdotes, separated by four decades, to underline the sharp contrast between medicine in the pre-*Belmont* days and medicine since then, in part as the result of the institutionalization of principles of respect for persons, beneficence, and justice in clinical medicine through a wide range of influences and forces. "In clinical medicine," Cassell contends, "the principles of the *Belmont Report* both influence and are influenced by the actions of individual patients and individual doctors and their relationships, even when the principles may be expressed by specific institutional procedures," such as institutional consent forms and court rulings about informed consent. Beyond institutional structures and requirements, "patterns of practice, professional ideals, and the everyday behavior of both doctors and patients . . . show what patients expect or demand and what physicians feel obligated to do."

The next several chapters focus directly on the three *Belmont* principles, examining their content and their possibilities and limitations for current and future ethical reflection, as well as identifying and discussing several unresolved questions. Karen Lebacqz, an ethicist at the Pacific School of Religion and a member of the National Commission, wonders whether various modifications and interpretations of the *Belmont* principles represent deeper wisdom or substantial loss. In particular, she challenges the emergence of the principle of respect for autonomy as a common but unfortunate narrowing of the principle of respect for persons. Such misinterpretations of the *Belmont* principles have produced "principlism," but justifiable criticisms of principlism do not obviate the need for principles in shaping public policy for research involving human subjects. Indeed, according to Lebacqz, the original *Belmont* principles remain valid as basic principles, but, she argues, we need to extend and reinterpret them in order "to retrieve their original depth and breadth and also to attend to subsequent developments in the refinement of understandings of oppression and liberation," as viewed through a feminist lens.

By contrast, Larry Churchill, a bioethicist at the Vanderbilt University School of Medicine, contends that the "principle of respect for persons" requires further revision in order to formulate a more robust principle of respect for autonomy. Instead of providing "a strong endorsement of respect for the autonomous choices of subjects, *Belmont's* understanding of 'respect for persons' presents a weak and distorted understanding of self-determination," which damages "the enterprise of research with human subjects." These problems, Churchill maintains, result from the *Belmont Report's* formulation, one that concentrates on protecting nonautonomous subjects and makes the investigator the primary moral agent who "gives weight to autonomous persons' considered opinions and choices." In so doing, the report fails to establish an obligation for researchers to promote and enable independent, autonomous choices, which are arguably even more important in research than in health care. Indeed, Churchill proposes dropping "respect for persons" as the first principle and relocating it in an introduction or preface where it can provide "the guiding vision for everything that follows." "Respect for persons" is best construed as "a basic or foundational commitment guiding research with human subjects, rather than as one of three principles accompanying 'beneficence' and 'justice.'" And "the essential first step" in respecting persons in research is to respect their autonomy.

The other two *Belmont* principles — beneficence and justice — also receive close scrutiny. Robert J. Levine, a member of the faculty of the Yale University School of Medicine and a consultant to the National Commission, concentrates on the principle of beneficence. Placing beneficence in the context of the National Commission's general view of principles, Levine notes that the *Belmont Report* does not provide a lexical ordering of the principles; that beneficence is not less important than respect for persons, for example; and that beneficence and the other principles are not direct action guides. Instead, norms, related to the principles, are action guides. The principle of beneficence "creates an obligation to secure the well-being of the individuals who serve as research subjects and to develop information that will form the basis of our being better able to serve the well-being of similar persons in the future." Levine then unpacks the interpretation of risk-benefit analysis and assessment required by beneficence, noting in particular the importance of the National Commission's rejection of the language of therapeutic and non-therapeutic research, which, however, still lingers and distorts risk-benefit judgments through what he calls "the fallacy of the package deal."

The next chapter, by Patricia King, another member of the National Commission and a professor at the Georgetown University Law Center, focuses on the principle of distributive justice — the distribution of benefits and burdens — and its application, in the *Belmont Report*, to the selection of subjects for participation in research. In King's formulation, the National Commission's conception of justice reflects its concern "to protect vulnerable research subjects from coercion and exploitation and to avoid a mismatch between those who were the subjects of research and those who were its primary beneficiaries." A major shift in perspective has occurred over the last twenty-six years, in part as a result of battles to gain access to experimental drugs in the HIV/AIDS epidemic. The result is an emphasis on "therapeutic research rather than nontherapeutic research, access rather than protection, and benefits rather than risks and burdens" of research. Whether the focus is on protection or access, the fair distribution of both benefits and burdens must attend to the relevance of different group memberships. Furthermore, as important as distributive justice is, King stresses the need for two conceptions of justice not addressed in the *Belmont Report*: procedural justice — for example, the participation of women and minorities in setting research agendas — and compensatory justice — for example, providing compensation to individuals for research-related injuries.

Continuing the focus on justice, philosopher Susan Sherwin of Dalhousie University in Nova Scotia, Canada, reexamines the *Belmont* principles through a "distinctly feminist lens," attending to "social justice" in the context of oppression of groups as well as of individuals. She notes that the *Belmont Report*, in passing, "anticipates many of the concerns that have troubled feminists in subsequent years," but does not really address them. By contrast, her feminist perspective begins with analysis of the nature of oppression and then examines how oppression differentially affects "differently situated social groups." The goal of her analysis is to gain "a better sense of what is needed to develop revised research ethics guidelines that more fully capture the demands of social justice." Such guidelines need to not only address research, as distinguished from therapy in the *Belmont Report*, but also cover innovative practice, in part because of its impact on women and other oppressed groups. According to Sherwin, *Belmont*'s examination of the risks of exploitation suffers from the lack of analysis of oppression: "What is missing is an analysis of oppression that explains how being a member of an oppressed group in society makes one particularly vulnerable to exploitation in research contexts. Such an account would explain why it is that membership in an oppressed group increases one's 'accessibility' to researchers." Sherwin further emphasizes the need for attention to the injustice of exclusion and underrepresentation as well as exploitation through inclusion and overrepresentation. In addition, "social justice" also concerns how the research agenda is set and includes the procedural requirement for "fair representation of diverse social groups throughout the full research process," as well as attention to the structures of oppression.

A communitarian perspective marks the chapter by Ezekiel Emanuel, a physician-bioethicist and chair of the Department of Clinical Bioethics of the National Institutes of Health (NIH), and Charles Weijer, a physician-bioethicist in the Department of Bioethics at Dalhousie University in Nova Scotia, Canada. Emanuel was a member of NBAC before joining NIH, and Weijer contributed to NBAC's deliberations as a consultant. In their chapter they consider how to locate the protection of communities in "the moral framework of research ethics" and what additional protections, if any, should be added. Using cases of psychiatric illness in an Amish community and social predictors of tamoxifen use among women at risk for breast cancer, Emanuel and Weijer note that, in addition to the risks for individual participants, such studies "also pose risks to the community as a whole," including, for instance, stigmatization and discrimination. Rather than trying to incorporate "community" into the other *Belmont* principles, which present an "individualistic vision," Emanuel and Weijer propose "a fourth principle for contemporary research ethics, a principle of respect for communities," as Emanuel proposed at NBAC's first meeting. Among their several pragmatic and substantive reasons for such an addition is the acknowledgment of the community's "moral status" as "more than the sum of individual values and interests; the community itself has values and interests" that are "worthy of protection." At a minimum, this new principle imposes on the researcher "an obligation to respect the values and interests of the community in research and, wherever possible, to protect the community from harm." The authors propose a five-step process for researchers and Institutional Review Boards (IRBs) to use in determining what kinds of protections are appropriate for what kinds of communities. Finally, conflicts may arise between communal choices and interests, on the one hand, and individual or group choices and interests, on the other; and while the authors stress that the *Belmont* principles should remain and provide a basis for criticizing some communal choices and interests, they do not indicate exactly how conflicts should be resolved.

Resolving conflicts among the *Belmont* principles, or among other bioethical principles, is the topic of the next two chapters, which consider how principles can be interpreted and applied. According to Robert Veatch, a faculty member of the Kennedy Institute of Ethics at Georgetown University and a consultant to the National Commission, the *Belmont* principles cannot always be applied because they may come into conflict. Indeed, this is "the single most crucial problem with *Belmont*: its failure to make clear what should happen when the principles conflict among themselves, that is, when, in the assessment of a proposed protocol, one principle supports one conclusion about a protocol while another supports a different conclusion." In particular, if the significant benefits of some research protocol could only be realized by compromising respect for persons or justice, what would the "moral theory" of the *Belmont Report* say? Three possibilities are (1) the simultaneity view, which holds that all the principles have to be satisfied simultaneously for a protocol to be ethically justified; (2) the balancing view, which requires that the principles be satisfied "on balance"; and (3) the ranking view, which claims that principles can be rank ordered, or lexically ordered, and that "the highest ranking principle must be fully satisfied before the next in rank is considered." While the "moral theory" of the *Belmont Report* is not wholly clear, Veatch's own approach incorporates elements of all three: together, components of the deontological principles of respect for persons, which he develops in an elaborate way, and of justice must be balanced, and together they are collectively ranked against beneficence (what he calls utility); they therefore must all then be satisfied simultaneously.

Further developing his own distinctive and influential approach to interpreting moral principles and resolving conflicts among them, Henry Richardson of the Department of Philosophy at Georgetown University considers whether and how bioethical principles can be specified, balanced, and interpreted. He argues against the metaphor of "balancing" in favor of a fuller development of specification, his preferred method, or interpretation. Richardson complains that "relying on the metaphor of balancing leads one to offer the mere semblance of reason giving, where real reason giving is wanted, and the mere appearance of guiding action, where actual guidance is wanted." After all, he notes, bioethical theory aims to guide actions. He is not opposed to one kind of balancing that is a feature or implication of norms — for example, balancing might be required to determine whether a risk is outweighed by probable benefits — but rather to balancing as a mode of conflict resolution (as seen in Beauchamp and Childress's *Principles of Biomedical Ethics*) and global or overall balancing (found, e.g., in the critiques by Gert, Culver, and Clouser). Specification, as one type of interpretation, is not an alternative "theory," but "an alternative way of conceiving of the relationship between a theory's norms and the guidance of action." In short, instead of balancing norms, we should specify or otherwise interpret them. Richardson does enter one caveat: "sometimes we may feel so tentative about the resolution we have reached in a concrete case that we feel we are not in a position to project it into the future. Sometimes we will do better just to admit that a problem needs further work. Notice, though, that in such instances, resting with balancing is being commended only weakly, as a second-best outcome."

The final chapter, written by John H. Evans, a sociologist at the University of San Diego, offers a sociological account of principlism through Max Weber's interpretation of the development of formal rationality. According to Evans, principlism, as represented by various works, including the *Belmont Report* and *Principles of Biomedical Ethics*, is a way to structure decision making in medicine and science through formally rational principles that provide calculability and predictability, especially where there is distrust of government. Evans suggests that this sociological perspective enables us better to explain why the principles were created and why principlism became dominant, to appreciate the attractiveness of several critiques of principlism and to make some predictions about the outcome of current debates about principlism, to sketch the benefits and costs of using moral principles, and to identify ways to minimize the costs of using principles. Those costs include the reduction of "thick" or complicated values into a "thin" or commensurable metric.

Childress's epilogue, "Looking Back to Look Forward," reflects further on some of the themes, including several unresolved issues, raised in these chapters in assessing the current and future role for the *Belmont* principles and other forms of principlism in biomedical research and practice.

Finally, we include the *Belmont Report* itself as an appendix to this volume. Just over 5,500 words long, this compact report, with its three principles, has been remarkably important and influential, even as the principles need regular attention and reinterpretation, if not modification and augmentation, as suggested by several chapters in this volume.

We gratefully thank each of the contributors to this volume as well as other conference participants, especially those who prepared papers that unfortunately could not be included because of considerations of length and those who served as discussants.

Note

1. Readers could be confused by seeing two different dates in this volume and elsewhere for the publication of the *Belmont Report*. Most references are to the *Belmont Report: Ethical Principles and Guidelines for the Protection of Human Subjects of Research*, *Federal Register*, April 18, 1979. However, the *Belmont Report* was also published earlier, in late 1978, as a separate volume: National Commission for the Protection of Human Subjects of Biomedical and Behavioral Research, *The Belmont Report: Ethical Principles and Guidelines for the Protection of Human Subjects of Research* (Washington, DC: DHEW Publication OS 78-0012), along with two appendices that provided the background papers considered by the National Commission.

PART I

Background and Origins of the *Belmont Report*

1 On the Origins and Future of the *Belmont Report*

Albert R. Jonsen

I was a member of the National Commission for the Protection of Human Subjects of Biomedical and Behavioral Research, which produced the *Belmont Report*. As such, I was a participant in the discussions that formed the substance of that report. In this chapter I describe the origins of this report, based on my recollections and the record, as best I can reconstruct both from the old lumber of my memory and the aging brick and mortar of old transcripts.[1] I will add to that account comments about influences that, in my opinion, shaped the National Commission's approach to the writing of *Belmont*, and then I will suggest directions toward its revision for the next bioethical era.

The congressional mandate required that the National Commission "identify the ethical principles which should underlie the conduct of biomedical and behavioral research with human subjects and develop guidelines that should be followed in such research." Of course, some statements of those principles already existed. The Nuremberg Code, devised in 1947 by the prosecuting team in the International War Crimes Trial, and the Declaration of Helsinki, issued by the World Medical Association in 1962, were prominent examples. However, the commissioners judged that they were being asked to explore the ethical foundations for human research more deeply than had any extant statements. They decided that a closed retreat should be held (meetings were usually open to the public by law) so that a freewheeling discussion could explore the nature and role of ethical principles for human research. That retreat was held at Belmont House, a conference center of the Smithsonian Institution in Elkridge, Maryland, February 13–16, 1976.

The commissioners, with a cadre of advisors, repaired to that pleasant eighteenth-century country house for a very twentieth-century debate. They had at hand a small library of essays prepared by scholars on the nature and role of moral principles in general and for research in particular. Other than a paper on research design from Alvin Feinstein, all the essays were by ethicists: Kurt Baier, Alasdair MacIntyre, James Childress, H. Tristram Engelhardt, and LeRoy Walters (we also had an earlier essay by Robert Veatch, which

3

linked the practice of informed consent to basic ethical principles). Staff philosopher Stephen Toulmin had prepared a meta-analysis of these essays, which he presented at the opening session. He stated that "in summary, the central question is how to reconcile protection of individual rights with fruitful pursuit of the collective enterprise."[2] After discussing Toulmin's presentation, the commissioners self-selected into small groups to discuss the principal elements of the congressional mandate: principles, risk-benefit, informed consent, and the boundaries between research and practice. After some six hours of group discussion, the general session reconvened, and the groups reported their conclusions.

The report of the ethical-principles group was delivered by Commissioner Karen Lebacqz. That group had selected seven principles: respect self-determination, benefit individual research subjects, benefit other individuals and groups present and future, minimize harm to individual subjects, minimize consequential harm to others, attend to distributive justice, and attend to compensating justice. Toulmin suggested that "protection of the weak and powerless" should be added. Guidelines for research based on these principles were good design, identification of consequences, informed consent, compensation, and selection of subjects. Early in the discussion, Commissioner Joseph V. Brady objected that the group had selected "too many principles and some of them, such as compensation, are not universal" (as Baier's essay had contended any true ethical principle must be). The list was not, said Brady, "crisp enough." He professed that he was attracted to three principles only, beneficence, freedom, and justice. I expressed agreement with Brady on the grounds of rhetorical brevity and the adequacy of those three principles for our task.

The commissioners had in their dossier of philosophical essays one by philosopher/physician H. Tristram Engelhardt. He suggested three basic principles: "respect for persons as free moral agents, concern to support the best interests of human subjects in research, intent to assure that the use of human subjects of experimentation will on the sum redound to the benefit of society." Philosopher Tom Beauchamp had contributed a paper titled "Distributive Justice and Morally Relevant Differences." The commissioners, after much discussion, took Engelhardt's first two principles and Beauchamp's principle of distributive justice and came up with the "crisp" principles that Brady and I desired: respect for persons, beneficence, and justice. A rough summation of the subcommittee reports had been drafted at the meeting. Stephen Toulmin was directed to redraft that document in light of the discussion and to present this version at the National Commission's March meeting.[3]

The *Belmont Report*, as the document came to be known, was the subject of long rumination after the Belmont meeting. Toulmin had drafted a version for the March 1976 meeting and had slightly modified that draft on June 6, 1976. Chairman Ryan described that draft as "a synthesis by Dr. Toulmin of the deliberations of Belmont."[4] Toulmin's two drafts, while circulated to the commissioners for study and comment, were not on the agenda for discussion until one year later, at the twenty-seventh meeting, February 11–13, 1977. In September, on the occasion of a National Commission meeting in my hometown, San Francisco, a small group of commissioners, consultants, and several staff persons convened in my study to revise the June 1976 draft in light of the February 1977 deliberations. We spent two days working on a text that would be succinct, easily comprehensible, and relevant to research practice. Three basic principles, "among those generally accepted in our cultural tradition, that are particularly relevant to the ethics of research involving human subjects: the principles of respect for persons, beneficence and justice," are defined. The application of these general principles leads to the three practical requirements of informed consent, risk-benefit assessment, and just selection of subjects for research.

This version was discussed at the January 13–14, 1978, meeting. By this time, the text, drafted and redrafted by many hands, had reached a form close to the final document. The discussion at that meeting made no substantial changes. A few refinements in language and, as Chairman Ryan said, some "cutting of fat," were done at a conference among myself, Toulmin, and the National Commission's new philosophy consultant, Tom Beauchamp, who was charged with producing a polished final version. The *Belmont Report* was approved by the commissioners at their forty-second meeting, on June 10, 1978. This short document, which was published in the *Federal Register* on April 18, 1979, not only provided an ethical framework for human experimentation but also had a major impact on the development of bioethics.[5] Its principles found their way into the general literature of the field, moving from the principles underlying the conduct of research to the basic principles of bioethics.[6]

The actual document, from conception to completion, was the work of many heads and hands. The intense discussions at Belmont House between commissioners, consultants, and staff initiated a long period of desultory discussion and revision, punctuated by concentrated redrafting. The several days in San Francisco revived the intense discussions of *Belmont* among a smaller circle and gave final shape to the document, which still needed polishing before the concluding debate and approval at a regular National Commission meeting. The final document was far from perfect. As I read it today, I see passages that are unclear and arguments that are unconvincing. However, given its purpose, it was a most satisfactory document. It was written to provide to the research community, including not only researchers but also administrators and regulators, a clear, concise statement of the principles that, if observed in particular practices, would render their enterprise ethical. *Belmont* was not to be a philosophical treatise nor was it to be a set of guidelines: it was a proclamation that had to ring true in the ears of scientists, policymakers, politicians, ethicists, journalists, and judges. It had to echo with commonly acknowledged public values rather than murmur arcane ethical arguments. Unquestionably, the sharp intellect of an analytic philosopher could find formulations that were clumsy or arguments that limped, but the document had to be both intelligible to those who were being called to live by its principles and sound enough to pass philosophical muster, even if it was not tuned to philosophical precision. Finally, *Belmont* was a document directed to one enterprise — scientific research involving human subjects — and to that enterprise as it existed and was understood at the time of its composition. It was not intended as a general exposition of the principles of bioethics or as a summary of principles that could be translated flawlessly into other areas of concern. It was a statement of the principles that should govern biomedical and behavioral research as understood at that time. For this particular purpose, it succeeds admirably.

Congress had instructed the National Commission to identify the ethical principles governing research with human subjects after a "comprehensive investigation." We could not merely invent them in our heads. It is not quite obvious what sort of investigation "identifies" ethical principles (*identify* is a peculiar word: it implies that the principles are lying out there somewhere and can be found by searching, like a treasure hunt). Should it be a philosophical inquiry, a sociological study, an opinion poll? Despite this uncertainty, the National Commission had to conduct some sort of investigation. The actual investigation, however, was rather informal. The explicit discussion of what was to become the *Belmont Report* did not begin until the second year of the National Commission's life. Yet, during those first years, we had constantly discussed and debated ethical principles as we argued our way through the other tasks set by Congress: research with the human fetus, with children, with incarcerated persons, and with institutionalized mentally infirm persons. The meeting at Belmont House was an opportunity to pause in those intense discussions and reflect on how our many references to principles might be formalized in a compact, comprehensive way. We also had learned, during the previous months of work, the history of the problems raised by research with human subjects. At the very first meeting of the National Commission, December 3, 1974, the commissioners were given copies of Jay Katz and Alex Capron's monumental *Experimentation with Human Beings* (1972), which led us into the labyrinths of that history. A comprehensive investigation also had to include inquiry into the thought of those familiar with the notion of ethical principles. The several commissioners who were familiar with the literature of philosophical and theological ethics were aware that the notion of ethical principles was not a settled and simple one, easily transferred from the pages of philosophy books into our recommendations. So we decided to invite five working ethicists to educate us on the meaning and use of that complex notion. We received from them the anthology of five excellent essays to which I referred earlier. Our comprehensive investigation also listened to the words of those ethicists who had already made a serious effort to identify the ethical principles governing research. Fortunately, two of the era's most outstanding ethical scholars had made such an attempt, and we had their work in our hands: Hans Jonas's tour de force "Philosophical reflections on experimenting with human subjects," written for the 1968 Academy of Arts and Sciences symposium on human experimentation, and the first chapter of Paul Ramsey's *Patient as Person* (1970), devoted to research with children.[7]

One could not read all this material without seeing one principle emerge dominant: the obligation to respect the autonomy of any person invited into research, with its corollary moral rule of informed consent. The many scholars who informed our study unanimously repudiated a utilitarian approach to the subject. Jonas did so quite explicitly when he criticized the words of Dr. Walsh McDermott, one of the nation's premier physicians, who had said, "The core of this ethical issue is to ensure the rights of society even if an arbitrary judgment must be made against an individual."[8] When Stephen Toulmin presented a meta-analysis of the scholarly essays on the first night at Belmont, he echoed McDermott's words, saying, "In summary, the central question is how to reconcile protection of individual rights with the fruitful pursuit of the collective enterprise." Yet, ironically, none of the scholars had done much reconciling at either the theoretical or practical level. Rather, they had come out loud and strong for the principle of autonomy and the protection of the subjects. The dominance of that principle is clearly expressed in some later words of Jay Katz:

> Had the Nuremberg Tribunal been aware of the tensions that have always existed between the claims of science and individual inviolability . . . it might have suggested that a balancing of these competing quests is necessary. . . . Even if the Tribunal had been aware of this problem, I hope it would not have modified its first principle, "The voluntary consent of the human subject is absolutely essential." It is this assertion that constitutes the significance of the Nuremberg Code then and now. Only when that principle is firmly put into practice can one address the claims of science and the wishes of society to benefit from science. Only then can one avoid the dangers that accompany a balancing of one principle against the other that assigns equal weight to both; for only if one gives primacy to consent can one exercise the requisite caution in situations where one may wish to make an exception to this principle but only for clear and sufficient reasons.[9]

The *Belmont Report* affirms that view: if research involving human persons as subjects is to be appraised as an ethical activity it must above all be an activity into which persons freely enter. This essential element was, we thought, captured within the broad scope of the idea of respect for persons.[10] The *Belmont Report* then added to respect for persons two other principles, beneficence and justice. Beneficence and its correlate, nonmaleficence, was an obvious addition, because all previous statements on the ethics of research, from Claude Bernard to Pope Pius XII, from Nuremberg to Helsinki, admonish the researcher not to harm the subject. Justice was less obvious, but its importance was suggested by the common but invidious practice of burdening the indigent sick with research whose beneficial products flowed to the better-off. Tuskegee was the shameful reminder of that practice. However, the larger question of the relationship between individual and society, raised by the words of McDermott and of Toulmin, which certainly can be framed as a question of justice, was addressed neither at Belmont nor in the *Belmont Report*. The National Commission proclaimed the three principles but did not attempt to articulate the balancing or priorities between these three. Respect for persons, beneficence, and justice are proclaimed the three formal principles, the pillars that uphold the ethics of research with human subjects. (I would have preferred the title "Belmont Proclamation" instead of *Belmont Report*, but the tone is too pompous and dictatorial.)

Yet, even without a specific articulation of rank, it is clear that the first of these principles dominates the other two. A careful reading of *Belmont* reveals that the manner in which beneficence and justice are discussed limits their meaning quite stringently to benefits and harms to individual subjects and justice in selection of individual or classes of subjects. Neither of these two principles manifests the broader meanings of which they are susceptible: beneficence does not here refer to utility or social good, and justice does not extend to the claims of a community over the individuals of which it is made up. *Belmont*, then, does what Katz imagines Nuremberg wished to do: it gives primacy to consent, a practice rooted in respect for the autonomy of persons.

The actual recommendations of the National Commission in the various areas of research involving children, the institutionalized mentally infirm, and prisoners are somewhat less adamantine. The commissioners believed as principlists; they worked as casuists. They saw all the principles as, in the jargon of moral philosophy, prima facie principles, generally ruling, but under rare and specific circumstances, allowing for exception. This is a respectable doctrine in moral philosophy, but it is perilous, both because the circumstances cannot be clearly previsioned and because unscrupulous persons are eager to discover exceptions to their own benefit. Still, even when exceptions are envisioned, as in the difficult Recommendation 6 of the *Report on Children as Research Subjects*, where more than minimal risk is presented to children who will not benefit in the context of a serious public health problem, such as an epidemic, they are built upon the principle of respect.

Have social or scientific circumstances rendered the three principles of *Belmont* obsolete? Do the three principles need augmentation? reformulation? Should certain trends in moral philosophy, such as communitarianism, dictate a rewrite that would, for example, locate respect for autonomy within a theory of social responsibility? Would a new *Belmont*, written from the perspective of feminist ethics, have a different color? My answer to these questions is no and yes. I believe the three principles should stand (leaving aside the intrabioethical disputes about whether beneficence and nonmaleficence should be divided or whether respect for autonomy should include protection of the nonautonomous). On the other hand, I believe that a new redaction of the text would be advisable. Allow me a metaphor to explain my ambiguous answer.

Belmont House, Elkridge, Maryland, a fine country mansion built in the late eighteenth century, was the site of the vigorous discussions that produced the *Belmont Report*. Early in 1999, I was invited to speak with the National Bioethics Advisory Commission (NBAC) about the revision of Belmont at the other end of the nation, in Portland, Oregon. That meeting of NBAC took place only a few miles from the terminus of the route traveled by Lewis and Clark when they explored the continent in 1804–05, a trip that took place but a few years after Belmont House was built. The Lewis and Clark expedition provides a metaphor for my suggested redaction of *Belmont*. The original report was drafted with an eastern seaboard perspective: a broad forested littoral sloping down to the Atlantic from the rugged but modest Appalachian range. *Belmont*'s perspective on the social and scientific enterprise called research was similarly flat and unspectacular. As Lewis and Clark labored westward, they were constantly amazed by the seemingly endless breadth of open prairie, the width and turbulence of the rivers, and, above all, the crowning height of the Rockies.

Twenty-five years of experience with the research enterprise has revealed similar dimensions of height, breadth, and width. We commonly refer to the AIDS experience. Only a few years after Belmont, the nation encountered an epidemic of communicable disease that many experts thought the civilized world would never see again. The epidemic conditions seemed to demand research, even at the price of individual autonomy. It also created a situation in which desperate persons demanded treatments as yet unproven and claimed a right to be research subjects. We have seen changes in the drug approval process to accommodate these demands. We have seen other epidemics, the appearance of other lethal viral diseases, and the recrudescence of resistant strains of tuberculosis and sexually transmitted diseases. Also research itself has expanded vastly as it moved out of the academic research centers into the pharmaceutical industry and feels the pressure of profit more than ever before. Research methodologies also have expanded. The controlled clinical trial remains the paradigm for research, but it has been crowded by all sorts of modifications to get at data difficult to enfold within classical protocols. It has become clear that research design systematically excluded women and that drugs used for pediatric conditions had not been tested in children: these two excluded classes of research subjects now must be added to those to whom the principle of justice in research must be applied. Treatment of persons with compromised mental capacity has been hindered by the difficulty of designing protocols that can ethically include them (the only National Commission report never implemented was *Research Involving the Institutionalized Mentally Infirm*). Development of genetic diagnosis challenges common notions of test accuracy and enters personal privacy more deeply than most biomedical research. Research involving whole communities of persons, particularly those with unique cultural heritages, must respect the customs of those groups. Research done abroad must both respect local norms and adhere to American standards. We now see more clearly than we did in 1976 that research and research ethics has its Rocky Mountains and its Columbia River: moral challenges that we did not see from the low, rolling plain of *Belmont*'s composition.

I believe that a redaction of *Belmont* for the next generation should retain almost unchanged the current text, what Biblical scholars would call the Ur-text. But the Ur-text should be surrounded by an appreciation of these broader, wider perspectives. I suggest that new frontiers can be delineated. The Ur-text contains three sections, A = Boundaries, B = Basic Ethical Principles, and C = Applications. I might now conceive of adding to each of these sections new redactions titled Frontiers, showing how the simple and straightforward or eastern seaboard perspective opens out into broader perspectives. First, the empirical frontiers, where classical scientific protocols meet other forms of investigation, should be described and their implications for ethical evaluation sketched. Second, the ethical frontiers, where the three principles meet and challenge each other: how should we describe the frontier, for example, between personal autonomy and social justice? Third, the frontiers where the scientific research enterprise encounters the demands of profit and of politics: how can we conceive of an ethical research enterprise under pressure from both? Without attempting the gargantuan task of exploring the new territories beyond the frontiers, some acknowledgment of their presence and immensity is desirable.

It is at the frontiers that serious ethical discourse and reasoning must be encouraged. Perhaps one of *Belmont*'s adverse effects was the impression that matters were settled: it came to be seen as the strict constructionists see the Constitution. I believe that a redaction should encourage the sense that, once principles are stated and their applications noted, the discussion has only begun. Ethics of research is a dynamic, casuistic activity. It is often said today that the excellent system of research review has stalled. May this not be, in part, because it became too automatic, too much the application of principles to protocols, and too little the struggle with the frontiers where principles confront previously unsuspected challenges?

I wish to affirm that my colleagues and I fully anticipated that an Ethical Advisory Board (EAB) would be established as a standing agency within the Department of Health and Human Services. We had so recommended in almost all of our reports. We expected that such a Board could be the living oracle of *Belmont*'s principles. Just as our Constitution requires a Supreme Court to interpret its majestically open-ended phrases, and, if I may allude to my own Catholic tradition, as the Bible requires a living Magisterium to interpret its mystic and metaphoric message, so does *Belmont*, a much more modest document than Constitution or Bible, require a constantly moving and creative interpretation and application. It was in the EAB that we envisioned the debate at the frontiers. It was from the EAB that we expected constant refreshing of the perspectives of members of Institutional Review Boards (IRBs) everywhere. It was to the EAB that we intended the apparently irreconcilable questions to go, if not for satisfactory resolution, at least for serious study and public exposition. None of that has taken place, as is generally well known. The EAB lingers in ghostly form, as an ignored imperative within 46.204 of the Federal Regulations 45 CFR 46. I earnestly hope that any revision of *Belmont* is matched by a revitalization of the EAB. So then, in my view, *Belmont* is an essentially sound proclamation. Its three principles are the right ones, necessary and sufficient for the ethics of research with human subjects. At the same time, those principles must illuminate wider territories, ethical and empirical, than they now do in the Ur-text. The written proclamation, whatever form it takes on paper, must be delivered to a body of responsible interpreters who can make its words come alive in the particular circumstances of particular protocols, public policy, and the changing research enterprise.

Notes

1. This account is adapted from my book, Albert R. Jonsen, *The Birth of Bioethics* (New York: Oxford University Press, 1998).

2. Transcript, 15th meeting, February 13–15, 1976, Meeting files, p. 4, National Commission Archive Box 26 [hereafter cited as Archive Box].

3. Transcript, 15th meeting, February 13–15, 1976, Meeting files, pp. 109–49, Archive Box 26.

4. Transcript, 27th meeting, February 11–13, 1977, Meeting files, p. 7, Archive Box 30.

5. *The Belmont Report: Ethical Principles and Guidelines for the Protection of Human Subjects of Research* (Washington, DC: U.S. Government Printing Office, 1978).

6. See Tom L. Beauchamp and James F. Childress, *Principles of Biomedical Ethics* (New York: Oxford University Press, 1979), now in its fifth edition (2001). The chapters of this important book were organized around four principles: autonomy, beneficence, nonmaleficence, and justice. Since it was being written at approximately the same time that both authors were working with the National Commission, a mutual influence of ideas was inevitable.

7. Jay Katz and Alexander M. Capron, *Experimentation with Human Subjects* (New York: Russell Sage Foundation, 1972); Hans Jonas, "Philosophical Reflections on Experimentation with Human Subjects," *Daedalus* 98 (1969): 219–47; Paul Ramsey, *Patient as Person* (New Haven: Yale University Press, 1970).

8. Jonas, "Philosophical Reflections," p. 230.

9. Jay Katz, "The Nuremberg Code Consent Principle: Then and Now," in *The Nazi Doctors and the Nuremberg Code*, ed. George Annas and Michael Grodin, 236–37 (New York and Oxford: Oxford University Press, 1992).

10. The phrase "respect for persons" was not widely employed in the technical vocabulary of ethics at the time of the National Commission's discussions. Engelhardt had used it in his essay for the National Commission. Commissioner Karen Lebacqz and I had taught a course during 1974 at Graduate Theological Union in which we used Robert Downie and Elizabeth Telfer, *Respect for Persons* (London: Allen and Unwin, 1969), which exposes the broad scope of Kant's concept of respect.

2 The Origins and Evolution of the *Belmont Report*

Tom L. Beauchamp

When, on December 22, 1976, I agreed to join the staff of the National Commission for the Protection of Human Subjects of Biomedical and Behavioral Research, my first and only major assignment was to write "The Belmont Paper," as it was then called. At the time, I had already drafted substantial parts of *Principles of Biomedical Ethics* with Jim Childress.[1] Subsequent to my appointment, the two manuscripts were drafted simultaneously, often side by side, the one inevitably influencing the other.

I here explain how the "Belmont Paper" evolved into the *Belmont Report*.[2] I will also correct some common but mistaken speculation about the emergence of frameworks of principles in research ethics and of the connections between *Belmont* and *Principles*.

The Beginnings of *Belmont*

The idea for the "Belmont Paper" originally grew from a vision of shared moral principles governing research that emerged during a break-out session at a four-day retreat held February 13–16, 1976, at the Smithsonian Institution's Belmont Conference Center in Maryland.[3] Earlier in this volume Albert Jonsen reports on the contributions of Stephen Toulmin, Karen Lebacqz, Joe Brady, and others. This meeting predates my work on the *Belmont Report*, and I leave it to Jonsen and others in attendance to relate these events.

A few months after this conference at Belmont, I received two phone calls, the first from Toulmin, staff philosopher at the National Commission, and the second from Michael Yesley, staff director. They asked me to write a paper for the National Commission on the nature and scope of the notion of justice. Yesley told me that the commissioners sought help in understanding theories of justice and their application to moral problems of human subject research. I wrote this paper and assumed that my work for the National Commission was concluded.[4]

However, shortly after I submitted the paper, Toulmin returned to full-time teaching at the University of Chicago, and Yesley inquired whether I was available to replace him on the staff. This appointment met some resistance. Two commissioners who later became my close friends — namely, Brady and Donald Seldin — were less than enthusiastic about my appointment. Nonetheless, Yesley prevailed, likely with the help of Chairperson Kenneth Ryan and my colleague Patricia King, and I joined the National Commission staff.

On my first morning in the office, Yesley told me that he was assigning me the task of writing the "Belmont Paper."[5] I asked Yesley what the task was. He pointed out that the National Commission had been charged by Congress to investigate the ethics of research and to *explore basic ethical principles*.[6] Members of the staff were at work on various topics in research ethics, he reported, but no one was working on basic principles. He said that an opening round of discussions of the principles had been held at the Belmont retreat. The National Commission had delineated a rough schema of three basic ethical principles: respect for persons, beneficence, and justice. I asked Yesley what these moral notions meant to the commissioners, to which he responded that he had no well-formed idea and that it was my job to figure out what the commissioners meant — or, more likely, to figure out what they should have meant.

So, I found myself with the job of giving shape and substance to something called the "Belmont Paper," though at that point I had never heard of Belmont or the paper. It struck me as an odd title for a publication. Moreover, this document had never been mentioned during my interview for the job or at any other time until Yesley gave me the assignment. My immediate sense was that I was the new kid on the block and had been given the dregs of National Commission work. I had thought, when I decided to join the National Commission staff, that I would be working on the ethics of psychosurgery and research involving children — heated and perplexing controversies at the time. I was chagrined to learn that I was to write something on which no one else was working and that had its origins in a retreat that I had not attended. Moreover, the mandate to do the work had its roots in a federal law that I had not until that morning seen.

Yesley proceeded to explain that no one had yet worked seriously on the sections of the report on *principles* because no one knew what to do with it. This moment of honesty was not heartening, but I was not discouraged either, because Childress and I were at that time well into the writing of our book on the role of basic principles in biomedical ethics. It intrigued me that we had worked relatively little on research ethics, which was the sole focus of the National Commission. I saw in my early conversations with Yesley that these two projects — *Principles* and *Belmont* — had many points of intersecting interest and could be mutually beneficial.

Yesley also gave me some hope by saying that a crude draft of the "Belmont Paper" already existed, though a twinkle in his eye warned me not to expect too much. That same morning I read the "Belmont draft."[7] Scarce could a new recruit have been more dismayed. So little was said about the principles that to call it a "draft" of principles is like calling a dictionary entry a scholarly treatise. Some sections were useful, especially a few pages that had been written largely by Robert Levine on the subject of "The Boundaries Between Biomedical and Behavioral Research and Accepted and Routine Practice" (later revised under the subtitle "Boundaries Between Practice and Research" and made the first section of the *Belmont Report*), but this draft of *Belmont* had almost nothing to say about the principles that were slated to be its heart.

In the next few weeks virtually everything in this draft pertaining to principles would be thrown away either because it contained too little on principles or because it had too much on peripheral issues. At the time, these peripheral issues constituted almost the entire document (with the exception of the section written by Levine, which was not peripheral, but also not on principles). The major topics addressed were the National Commission's mandate, appropriate review mechanisms, compensation for injury, national and international regulations and codes, research design, and other items that did not belong in the "Belmont Paper." These topics, being peripheral, were therefore eliminated. Except for Levine's section on boundaries, everything in this draft landed on the cutting-room floor.[8]

Once the Belmont "draft" was left with nothing in the section on principles, Yesley suggested that I might find the needed content from the massive compendium on research titled *Experimentation with Human Beings*, edited by Jay Katz with the assistance of Alexander Capron and Eleanor Swift Glass.[9] Drawn from sociology, psychology, medicine, and law, this book was at the time the most thorough collection of materials on research ethics and law. Yesley informed me that I should endeavor to learn all the information presented in this book, but after days of poring over this rich resource, I found that it offered virtually nothing on *principles* suitable for an analytical discussion of research ethics. The various codes and statements by professional associations found in this book had occasional connections with my task and with the National Commission's objectives, but only distant ones.[10]

The Historical Origins of the Principles of the *Belmont Report*

Fortunately, Childress and I had gathered a useful collection of materials on principles and theories, largely in the writings of philosophers. I had been influenced by the writings of W. D. Ross and William Frankena. My training led me to turn to these and other philosophical treatments, which had already proved helpful in my work on *Principles*.

However, it would be misleading to suggest that the principles featured in the *Belmont Report* derived from the writings of philosophers. Their grounding is ultimately in what I would eventually call (following Alan Donagan) "the common morality." The *Belmont Report* makes reference to "our cultural tradition" as the basis of its principles, and it is clear that these principles derive from the common morality rather than a particular philosophical work or tradition. However, what *Belmont* means by our "tradition" is unclear, and I believe the import of the *Belmont* principles cannot be tied to a particular tradition, but rather to a conviction that there is a universally valid point of view. I believe, and I think that the commissioners believed, that these principles are norms shared by all morally decent persons.

The commissioners almost certainly believed that these principles are already embedded in public morality and are presupposed in the formulation of public and institutional policies. The principles do not deviate from what every morally sensitive person knows to be right, based on their own moral training and experience. That is, every morally sensitive person believes that a moral way of life requires that we respect persons and take into account their well-being in our actions. *Belmont*'s principles are so woven into the fabric of morality in morally sensitive cultures that no responsible research investigator could conduct research without reference to them.[11]

The Relationship Between the *Belmont* Principles and the Principles in *Principles of Biomedical Ethics*

Many have supposed that the *Belmont Report* provided the starting point and the abstract framework for *Principles of Biomedical Ethics*.[12] They have wrongly assumed that *Belmont* preceded and grounded *Principles*.[13] The two works were written simultaneously, the one inevitably influencing the other.[14] There was reciprocity in the drafting, and influence ran bilaterally. I was often simultaneously drafting material on the same principle or topic both for the National Commission and for my colleague Childress, while he was at the same time writing material for me to inspect. I would routinely write parts of the *Belmont Report* during the day at the National Commission headquarters on Westbard Avenue in Bethesda, then go to my office at Georgetown in the evening and draft parts of chapters for Childress to review. Despite their entirely independent origins, these projects grew up and matured together.

Once I grasped the moral vision of the National Commission initiated at Belmont, I could see that Childress and I had major substantive disagreements with the National Commission. The names of the principles articulated in the *Belmont Report* bear notable similarities to some of the names Childress and I were using and continued to use, but the two schemas of principles are far from constituting a uniform name, number, or conception. Indeed, the two frameworks are not consistent.

I thought at the time, and still do, that the National Commission was confused in the way it delineated the principle of respect for persons. It seemed to blend two independent principles: a principle of respect for autonomy and a principle of protecting and avoiding the causation of harm to incompetent persons. Furthermore, Childress and I both thought that we should stick to our thesis that the principle of beneficence must be distinguished from the principle of nonmaleficence, though the National Commission failed to make any such distinction. This matter was connected to another problem that later bothered me, namely, that the National Commission had an all too utilitarian vision of beneficence—one with inadequate internal controls in its moral framework to protect subjects against abuse when there was the promise of major benefit for other sick or disabled persons.

The differences between the philosophy in *Principles* and the National Commission's views in *Belmont* have occasionally been the subject of published commentary.[15] Some commentators correctly see that we developed substantially different moral visions and that neither approach was erected on the foundations of the other. By early 1977, I had come to the view that the National Commission, especially in the person of its chair, Kenneth Ryan, was sufficiently rigid in its vision of the principles that there was no way to substantially alter the National Commission's conception, although I thought that all three principles were either defective or underanalyzed. From this point forward, I attempted to analyze principles for the National Commission exclusively as I thought the commissioners would find acceptable. *Principles of Biomedical Ethics* became the only work that reflected my own deepest philosophical convictions about principles.

The Drafting and Redrafting of *Belmont*

While Yesley gave me free rein in the drafting and redrafting of *Belmont*, the drafts were always subject to revisions and improvements made by the commissioners and staff members.[16] All members of the staff did their best to formulate ideas that were responsive to changes suggested by the commissioners. Commissioner Seldin encouraged me with as much vigor as he could muster (which was—and remains today—considerable) to make my drafts as philosophical as possible. Seldin wanted some Mill here, some Kant there, and the signature of philosophical argument sprinkled throughout the document. I tried this style, but other commissioners wanted a streamlined document and minimalist statement relatively free of the trappings of philosophy. Seldin, Yesley, and I ultimately relented,[17] and bolder philosophical defenses of the principles were gradually stripped from the body of *Belmont*.

Public deliberations in National Commission meetings were a staple source of ideas, but a few commissioners spoke privately to me or to Yesley about desired changes, and a few commissioners proposed changes to Assistant Staff Director Barbara Mishkin, who passed them on to me. Most of these suggestions were accepted, and a serious attempt was made to implement them. In this respect, the writing of this document was a joint product of commissioner-staff interactions. However, most of the revisions made by the commissioners (other than through their comments in public deliberations) concerned small matters, and the commissioners were rarely involved in making written changes.

One meeting on *Belmont* involving a few members of the staff and a few commissioners occurred in September 1977 in the belvedere — rooftop study — of Jonsen's home in San Francisco. The small group in attendance attempted to revise the "Belmont Paper" for presentation at the next meetings during which the commissioners were scheduled to debate it. As Jonsen reports in *The Birth of Bioethics* and in the present volume, the purpose of this meeting was to revisit previous drafts and deliberations of the commissioners.[18]

The history of drafting and redrafting that I have outlined may suggest that the document grew in size over time, but the reverse is true. The document grew quickly in the early drafting and then was contracted over time. I wrote much more for the National Commission about respect for persons, beneficence, and justice than eventually found its way into the *Belmont Report*. When various materials I had written were eliminated from *Belmont*, I would scoop up the reject piles and fashion them for *Principles of Biomedical Ethics*. Several late-written chunks of this book on research ethics were fashioned from the more philosophical, but abjured parts of what I wrote for the National Commission that never found its way into the final draft of *Belmont*.

Explicit and Implicit Ideas about an Applied Research Ethics

Michael Yesley deserves credit for one key organizing conception in this report. He and I were in almost daily discussion about the "Belmont Paper." We spent many hours discussing the best way to develop the principles, to express what the commissioners wanted to say, and even how to sneak in certain lines of thought that the commissioners might not notice. One late afternoon we were discussing the overall enterprise. We discussed each principle, whether the principles were truly independent, and how the principles related to the topics in research ethics under consideration by the National Commission. Yesley said, as a way of summarizing our reflections, "What these principles come to is really quite simple for our purposes: Respect for persons applies to informed consent, beneficence applies to risk-benefit assessment, and justice applies to the selection of subjects." Yesley had articulated the following abstract schema:

Principle of	*Applies to*	*Guidelines for*
Respect for Persons		Informed Consent
Beneficence		Risk-Benefit Assessment
Justice		Selection of Subjects

This schema may seem trifling; certainly it was already nascent in preexisting drafts of the report and in the National Commission's deliberations. But no one at the time had articulated the schema in precisely this way, and Yesley's summary was immensely helpful in peering through countless hours of discussion to see the underlying structure and commitment at work in the principles destined to be the backbone of *Belmont*. Yesley had captured what would soon become the major portion of the table of contents of the *Belmont Report*, as well as the rationale of its organization. I then attempted to draft the document so that the basic principles could be "applied" to develop guidelines in specific areas and could also serve as justification for guidelines.

In light of this schema, a general strategy emerged for handling problems of research ethics, namely, that each principle made moral demands in a specific domain of responsibility for research. For example, the principle of respect for persons demands informed and voluntary consent. Under this conception, the *purpose* of consent provisions is not protection from risk, as many earlier federal policies seemed to imply, but the protection of autonomy and personal dignity, including the personal dignity of incompetent persons incapable of acting autonomously, for whose involvement a duly authorized third party must consent.

I wrote the sections on principles in the *Belmont Report* based on this model of each principle applying to a zone of moral concern. In this drafting, the focus of the document shifted to include not only abstract principles and their analysis but also a moral view that is considerably more concrete and meaningful for those engaged in the practice of research. Explication of the values being advanced was heavily influenced by the context of biomedicine (and rather less influenced by contexts of the social and behavioral sciences). *Belmont*, in this way, moved toward an applied, professional morality of research ethics.

Although *Belmont* takes this modest step in the direction of an applied research ethics, there was never any ambition or attempt to make this document specific and practical. This objective was to be accomplished by the other volumes the National Commission issued. *Belmont* was meant to be, and should be remembered as, a moral framework for research ethics. Commissioners and staff were always aware that this framework is too indeterminate *by itself* to decide practice or policy or to resolve moral conflicts. The process of making the general principles in *Belmont* sufficiently concrete is a progressive process of reducing the indeterminateness and abstractness of the general principles to give them increased action-guiding capacity.[19] *Belmont* looks to educational institutions, professional associations, government agencies, and the like to provide the particulars of research ethics.

Principlism, Casuistry, or Both?

The final editing of the *Belmont Report* was done by three people in a small classroom at NIH.[20] Al Jonsen, Stephen Toulmin, and I were given this assignment by the National Commission. Some who have followed the later writings of Jonsen and Toulmin on casuistry may be surprised to learn that throughout the National Commission's deliberations, as well as in this final drafting, Jonsen and Toulmin contributed to the clarification of the *principles* in the report. There was never an objection by either that a strategy of using principles should be other than central to the National Commission's statement of its ethical framework. Jonsen repeats his support for these principles in the present volume.

However, Jonsen also says in this volume, "The commissioners believed as principlists; they worked as casuists." He is suggesting that the National Commission's deliberations constituted a casuistry of reasoning about historical and contemporary cases,[21] despite the commissioners' commitment to and frequent reference to *Belmont* principles. Jonsen and Toulmin once explicated this understanding of the National Commission's work as follows:

> The one thing [*individual* commissioners] could not agree on was *why* they agreed. . . . Instead of securely established universal principles, . . . giving them intellectual grounding for particular judgments about specific kinds of cases, it was the other way around.
>
> The *locus of certitude* in the Commissioners' discussions . . . lay in a shared perception of what was specifically at stake in particular kinds of human situations. . . . That could never have been derived from the supposed theoretical certainty of the principles to which individual Commissioners appealed in their personal accounts.[22]

This interpretation gives insight into the National Commission, but it needs careful qualification to avoid misunderstanding. Casuistical reasoning more so than moral theory or universal abstraction often did function to forge agreement during National Commission deliberations. The commissioners appealed to particular cases and families of cases, and consensus was reached through agreement on cases and generalization from cases when agreement on an underlying theoretical rationale would have been impossible.[23] Commissioners would never have been able to agree on a single ethical theory, nor did they even attempt to buttress the *Belmont* principles with a theory. Jonsen and Toulmin's treatment of the National Commission is, in this regard, entirely reasonable, and a similar line of argument can be taken to explicate the methods of reasoning at work in other bioethics commissions.[24]

Nonetheless, this methodological appraisal is consistent with a firm commitment to moral principles; the commissioners, including Jonsen, were emphatic in their support of and appeals to the general moral principles delineated in the *Belmont Report*.[25] The transcripts of the National Commission's deliberations show a constant movement from principle to case, and from case to principle. Principles supported argument about how to handle a case, and precedent cases supported the importance of commitment to principles. Cases or examples favorable to one point of view were brought forward, and counterexamples then advanced. Principles were invoked to justify the choice and use of both examples and counterexamples. On many occasions an argument was offered that a case judgment was irrelevant or immoral in light of the commitments of a principle.[26] The National Commission's deliberations and conclusions are best understood in terms of reasoning in which principles are interpreted and specified by the force of examples and counterexamples that emerge from experience with cases.

It is doubtful that Jonsen ever intended to deny this understanding of principles and their roles, despite the widely held view that casuistry dispenses with principles. Jonsen has said that "casuistic analysis does not deny the relevance of principle and theory,"[27] and, in an insightful statement in his later work, he has written that

> when maxims such as "Do no harm," or "Informed consent is obligatory," are invoked, they represent, as it were, cut-down versions of the major principles relevant to the topic, such as beneficence and autonomy, cut down to fit the nature of the topic and the kinds of circumstances that pertain to it.[28]

Jonsen goes on to point out that casuistry is "complementary to principles" and that "casuistry is not an alternative to principles: No sound casuistry can dispense with principles."[29]

Casuists and those who support frameworks of principles like those in *Belmont* and *Principles* should be able to agree that when they reflect on cases and policies, they rarely have in hand *either* principles that were formulated without reference to experience with cases *or* paradigm cases lacking a prior commitment to general norms. Only a false dilemma makes us choose between the National Commission as principlist or casuist. It was both.

Notes

This chapter was written for a conference on the *Belmont Report* held in Charlottesville, Virginia. It draws in part on my chapter in *The Story of Bioethics*, published by Georgetown University Press, 2003.

1. A contract for the book was issued by the Oxford University Press on August 19, 1976. The manuscript was completed in late 1977; galleys arrived in October 1978, bearing the 1979 copyright date.

2. See also *Appendices I and II* to the *Belmont Report.* The *Belmont Report* was completed in late 1977 and published on September 30, 1978. *The Belmont Report: Ethical Guidelines for the Protection of Human Subjects of Research* (Washington, DC: DHEW Publication OS 78-0012). It first appeared in the *Federal Register* on April 18, 1979.

The National Commission for the Protection of Human Subjects of Biomedical and Behavioral Research was established July 12, 1974, under the National Research Act, Public Law 93-348, Title II. The first meeting was held December 3–4, 1974. The 43rd and final meeting was on September 8, 1978.

3. See the archives of the National Commission, 15th meeting, February 13–16, 1976, vol. 15A — a volume prepared for the Belmont meeting. This meeting book contains a "staff summary" on the subject of "ethical principles" as well as expert papers prepared by Kurt Baier, Alasdair MacIntyre, James Childress, H. Tristram Engelhardt, Alvan Feinstein, and LeRoy Walters. The papers by Engelhardt and Walters most closely approximate the moral considerations ultimately treated in the "Belmont Paper," but neither quite matches the National Com-

mission's three principles. Walters, however, comes very close to a formulation of the concerns in practical ethics to which the National Commission *applies* its principles. All meeting books are housed in the archives of the Kennedy Institute of Ethics Library, storage facility, Georgetown University.

4. The paper was published as "Distributive Justice and Morally Relevant Differences" in *Appendix I to the Belmont Report*, pp. 6.1–6.20. This paper was distributed at the 22nd meeting of the National Commission, held in September 1976, seven months after the retreat at the Belmont Conference House.

5. My first day was the Saturday meeting of the National Commission on January 8, 1977. Yesley and I met the following Monday.

6. The National Research Act, P.L. 93-348, July 12, 1974. Congress charged the National Commission with recommending regulations to the Department of Health, Education, and Welfare (DHEW) to protect the rights of research subjects and developing ethical principles to govern the conduct of research. In this respect, the *Belmont Report* was at the core of the tasks the National Commission had been assigned by Congress. DHEW's conversion of its grants administration *policies* governing the conduct of research involving human subjects into formal *regulations* applicable to the entire department was relevant to the creation of the National Commission. In the U.S. Senate, Senator Edward Kennedy, with Jacob Javits's support, was calling for a permanent, *regulatory* commission independent of the National Institutes of Health (NIH) to protect the welfare and rights of human subjects. Paul Rogers in the House supported NIH in advocating that the commission be *advisory* only. Kennedy agreed to yield to Rogers if DHEW published satisfactory regulations. This compromise was accepted. Regulations were published on May 30, 1974; then, on July 12, 1974, P.L. 93-348 was modified to authorize the National Commission as an advisory body. Charles McCarthy helped me understand this history. For a useful framing of the more general regulatory history, see Joseph V. Brady and Albert R. Jonsen (two commissioners), "The Evolution of Regulatory Influences on Research with Human Subjects," in *Human Subjects Research*, ed. Robert Greenwald, Mary Kay Ryan, and James E. Mulvihill, 3–18 (New York: Plenum Press, 1982).

7. This draft had a history beginning with the 16th meeting (March 12–14, 1976), which contained a draft dated March 1, 1976, and titled "Identification of Basic Ethical Principles." This document summarized the relevant historical background and set forth the three "underlying ethical principles" that came to form the National Commission's framework. Each principle was discussed in a single paragraph. This document was slightly recast in a draft of June 3, 1976 (prepared for the 19th meeting, June 11–13, 1976), in which the discussion of principles was shortened to little more than one page devoted to all three principles. Surprisingly, in the summary statement (p. 9), "respect for persons" is presented as the principle of "autonomy." No further draft is presented to the National Commission until ten months later, at the 29th meeting (April 8–9, 1977). I began work on the document in January 1977.

Transcripts of the National Commission's meetings are also available in the archives of the Kennedy Institute of Ethics at Georgetown University. See

National Commission for the Protection of Human Subjects of Biomedical and Behavioral Research. Archived Materials 1974–78, General Category: "Transcript of the Meeting Proceedings" (for discussion of the "Belmont Paper" at the following meetings: February 11–13, 1977; July 8–9, 1977; April 14–15, 1978; and June 9–10, 1978).

8. Cf. the radical differences between the draft available at the 19th meeting (June 11–13, 1976) and the draft at the 29th meeting (April 8–9, 1977). The drafts show that the critical period that gave shape to the *Belmont* principles was the period between January and April 1977. Less dramatic improvements were made between April 1977 and eventual publication more than a year later.

9. Jay Katz, with the assistance of Alexander Capron and Eleanor Glass, eds., *Experimentation with Human Beings* (New York: Russell Sage Foundation, 1972).

10. The first and only footnote in the *Belmont Report* is a reference to this background reading. Typical materials that I examined during this period include *United States* v. *Karl Brandt, Trials of War Criminals Before the Nuremberg Military Tribunals Under Control Council Law No. 10*, 1948–49, Military Tribunal I (Washington, DC: U.S. Government Printing Office, 1948–1949), vols. 1 and 2, reproduced in part in Katz, *Experimentation with Human Beings*, 292–306; American Medical Association, House of Delegates, Judicial Council, "Supplementary Report of the Judicial Council," *Journal of the American Medical Association* 132 (1946): 90; World Health Organization, 18th World Medical Assembly, Helsinki, Finland, "Declaration of Helsinki: Recommendations Guiding Medical Doctors in Biomedical Research Involving Human Subjects," *New England Journal of Medicine* 271 (1964): 473, reprinted in Katz, *Experimentation with Human Beings*, 312–13. Less helpful than I had hoped was Henry Beecher, "Some Guiding Principles for Clinical Investigation," *Journal of the American Medical Association* 195 (1966): 1135–36. For behavioral research, I started with Stuart E. Golann, "Ethical Standards for Psychology: Development and Revisions, 1938–1968," *Annals of the New York Academy of Sciences* 169 (1970): 398–405, and American Psychological Association, Inc., *Ethical Principles in the Conduct of Research with Human Participants* (Washington, DC: APA, 1973).

11. For a clearer presentation of this viewpoint in a later document by a government-initiated commission, see several chapters in Advisory Committee on Human Radiation Experiments (ACHRE), *Final Report of the Advisory Committee on Human Radiation Experiments* (New York: Oxford University Press, 1996).

12. See, e.g., Eric Meslin et al., "Principlism and the Ethical Appraisal of Clinical Trials," *Bioethics* 9 (1995): 399–403; Bernard Gert, Charles M. Culver, and K. Danner Clouser, *Bioethics: A Return to Fundamentals* (New York: Oxford University Press, 1997), 72–74; and Jonathan D. Moreno, *Deciding Together: Bioethics and Consensus* (New York: Oxford University Press, 1995), 76–78. Meslin et al. say that "Beauchamp and Childress's *Principles of Biomedical Ethics* . . . is the most rigorous presentation of the principles initially described in the *Belmont Report*" (p. 403). Gert et al. see *Principles* as having "emerged from the work of the National Commission" (p. 73). Moreno presumes that Beauchamp and Childress "brought the three [*Belmont*] principles into bioethical analysis more gen-

erally." The thesis that the idea of an abstract framework of principles for bioethics originated with the National Commission has been sufficiently prevalent that authors and lecturers have occasionally cited the principles as Childress and I have named and articulated them, and then felt comfortable in attributing these same principles to the National Commission.

13. The draft of *Belmont* that appeared in typescript for the National Commission meeting of December 2, 1977 (37th meeting) shows several similarities to various passages in the first edition of *Principles of Biomedical Ethics*. Childress and I had completed our manuscript by this date, but *Belmont* would be taken through five more drafts presented to the commissioners, the last being presented at the final (43rd) meeting (September 8, 1978).

14. Prior to my involvement with the National Commission, and prior to the Belmont retreat, Childress and I had lectured and written about principles of biomedical ethics. In early 1976, coincidentally at about the same time of the Belmont retreat, Childress and I wrote a programmatic idea for the book (based on our lectures), which we submitted to the Oxford University Press. We had already developed a general conception of what later came to be called by some commentators "mid-level principles."

For more on the nature, history, and defensibility of a commitment to such mid-level principles in bioethics, see Gert, Culver, and Clouser, *Bioethics: A Return to Fundamentals*, 72ff; Beauchamp and David DeGrazia, "Principlism," in *Bioethics: A Philosophical Overview*, vol. 1 of *Handbook of the Philosophy of Medicine*, ed. George Khusfh (Dordrecht, Neth.: Kluwer, 2002); James Childress, "Ethical Theories, Principles, and Casuistry in Bioethics: An Interpretation and Defense of Principlism," in *Religious Methods and Resources in Bioethics*, ed. Paul F. Camenisch, 181–201 (Boston: Kluwer, 1994); Tom Beauchamp, "Principles and Other Emerging Paradigms for Bioethics," *Indiana Law Journal* 69 (1994): 1–17; Beauchamp, "The Four Principles Approach to Medical Ethics," in *Principles of Health Care Ethics*, ed. R. Gillon, 3–12 (London: John Wiley & Sons, 1994); Beauchamp, "The Role of Principles in Practical Ethics," in *Philosophical Perspectives on Bioethics*, ed. L. W. Sumner and J. Boyle (Toronto: University of Toronto Press, 1996); and Earl Winkler, "Moral Philosophy and Bioethics: Contextualism versus the Paradigm Theory," in *Philosophical Perspectives on Bioethics*, ed. L. W. Sumner and J. Boyle (Toronto: University of Toronto Press, 1996).

15. Particularly insightful is Ernest Marshall, "Does the Moral Philosophy of the *Belmont Report* Rest on a Mistake?" *IRB: A Review of Human Subjects Research* 8 (1986): 5–6. See also John Fletcher, "Abortion Politics, Science, and Research Ethics: Take Down the Wall of Separation," *Journal of Contemporary Health Law and Policy* 8 (1992): 95–121, esp. sect. 4, "Resources in Research Ethics: Adequacy of the *Belmont Report*."

16. Albert Jonsen, a commissioner, reports in *The Birth of Bioethics* (New York: Oxford University Press, 1998) that I was "working with [Stephen] Toulmin on subsequent drafts" of the *Belmont Report* in 1977. Although I sat next to Stephen in National Commission meetings and conversed with him about many subjects during meetings of the National Commission throughout 1977, we

never jointly drafted, worked on, or discussed *Belmont* until it was in near final form and already approved by the commissioners. Stephen had been assigned to a project on recombinant DNA and did not participate in *Belmont* drafts after I came to the National Commission.

17. Yesley was my constant critic, more so than anyone else. Seldin was my constant counsel, forever exhorting me to make the document more philosophically credible. Patricia King taught me more about the National Commission and its commissioners than anyone else. It was she who helped me understand why a really philosophical document was not the most desirable result.

18. Jonsen reports in *The Birth of Bioethics* (p. 103) that "the date is uncertain" of this meeting at his home, but the date is the afternoon of September 21 through September 23, 1977 (including travel period). Jonsen correctly remarks that one purpose of this meeting was to revisit "the February 1977 deliberations" of the commissioners, but he incorrectly reports that the meeting was called "to revise the June 1976 draft." Except for a section on boundaries written by Robert Levine, the June 1976 draft had been so heavily recast that *Belmont* was by September a completely different document. Seven months of continual redrafting of *Belmont* had occurred (from February to September 1977) prior to the meeting at his home. The National Commission did not discuss the draft reports at its meetings during those seven months (the last discussion having occurred in February). However, "staff drafts" of the "Belmont Paper" were distributed at two meetings during this period, namely, the 29th meeting (April 8–9, 1977) and the 30th meeting (May 13–14, 1977). All drafts are now housed in the archives of the Kennedy Institute of Ethics Library.

19. See Henry S. Richardson, "Specifying Norms as a Way to Resolve Concrete Ethical Problems," *Philosophy and Public Affairs* 19 (1990): 279–310; Richardson, "Specifying, Balancing, and Interpreting Bioethical Principles," *Journal of Medicine and Philosophy* 25 (2000): 285–307, a version of which appears in this volume; Tom Beauchamp and James Childress, *Principles of Biomedical Ethics*, 5th ed. (New York: Oxford University Press, 2001), 15–19; David DeGrazia, "Moving Forward in Bioethical Theory: Theories, Cases, and Specified Principlism," *Journal of Medicine and Philosophy* 17 (1992): 511–39; DeGrazia and Beauchamp, "Philosophical Foundations and Philosophical Methods," in *Methods in Medical Ethics*, ed. Jeremy Sugarman and Daniel P. Sulmasy (Washington, DC: Georgetown University Press, 2001), esp. 31–46.

20. I have diary notes that our editing meetings occurred May 31–June 1, 1978, just prior to the National Commission's 42nd and penultimate meeting, on June 9–10, 1978. Thus, the wording at the 42nd meeting was the final wording unless a commissioner raised an objection at that meeting or the next.

21. As used here, *casuistry* implies that some forms of moral reasoning and judgment neither appeal to nor rely upon principles and rules, but rather involve appeals to the grounding of moral judgment in narratives, paradigm cases, analogies, models, classification schemes, and even immediate intuition and discerning insight. Each change in the circumstances changes the case. The casuistic method begins with cases whose moral features and conclusions have already

been decided and then compares the salient features in the paradigm cases with the features of cases that require a decision. See Albert Jonsen and Stephen Toulmin, *Abuse of Casuistry* (Berkeley: University of California Press, 1988), 11–19, 251–54, 296–99; Jonsen, "Casuistry as Methodology in Clinical Ethics," *Theoretical Medicine* 12 (1991): 299–302; John Arras, "Principles and Particularity: The Role of Cases in Bioethics," *Indiana Law Journal* 69 (1994): 983–1014. See also Toulmin, "The Tyranny of Principles," *Hastings Center Report* 11 (1981): 31–39.

22. Jonsen and Toulmin, *Abuse of Casuistry*, 16–19.

23. A few years ago, I reviewed all the National Commission transcripts pertaining to *Belmont*, primarily to study the National Commission's method of treating issues in research ethics. I found that Jonsen and Toulmin had occasionally mentioned casuistry, but they clearly understood the National Commission's casuistry as consistent with its invocation of moral principles. For additional discussion, see Stephen Toulmin, "The National Commission on Human Experimentation: Procedures and Outcomes," in *Scientific Controversies: Case Studies in the Resolution and Closure of Disputes in Science and Technology*, ed. H. T. Engelhardt Jr. and A. Caplan, 599–613 (New York: Cambridge University Press, 1987), and Jonsen, "Casuistry as Methodology in Clinical Ethics."

24. See, e.g., Alexander Capron, "Looking Back at the President's Commission," *Hastings Center Report* 13, no. 5 (1983): 8–9.

25. See Jonsen's own summation to this effect in "Casuistry," in *Methods of Bioethics*, ed. Sugarman and Sulmasy, pp. 112–13, and his Introduction to a reprinting of the *Belmont Report* in *Source Book in Bioethics*, ed. Albert Jonsen, Robert M. Veatch, and LeRoy Walters, 22 (Washington, DC: Georgetown University Press, 1998).

26. See, e.g., National Commission for the Protection of Human Subjects of Biomedical and Behavioral Research, "Transcript of the Meeting Proceedings" for the following meetings: February 11–13, 1977, 11–155; July 8–9, 1977, 104–17; April 14–15, 1978, 155–62; and June 9–10, 1978, 113–19.

27. Albert Jonsen, "Case Analysis in Clinical Ethics," *The Journal of Clinical Ethics* 1 (1990): 65.

28. Albert Jonsen, "Casuistry: An Alternative or Complement to Principles?" *Kennedy Institute of Ethics Journal* 5 (1995): 237–51.

29. Jonsen, "Casuistry: An Alternative or Complement to Principles?" 244–49. See also Jonsen, "The Weight and Weighing of Ethical Principles," in *The Ethics of Research Involving Human Subjects: Facing the 21st Century*, ed. Harold Y. Vanderpool, 59–82 (Frederick, MD: University Publishing Group, 1996).

PART II

The *Belmont* Principles: Influence and Application

3 The Dog in the Night-Time

Or, The Curious Relationship of the *Belmont Report* and the President's Commission

Alexander M. Capron

"Is there any other point to which you would wish to draw my attention?"
"To the curious incident of the dog in the night-time."
"The dog did nothing in the night-time."
"That was the curious incident," remarked Sherlock Holmes.
—Sir Arthur Conan Doyle, *Silver Blaze*

At the moment of this exchange with Inspector Gregory, Holmes—who had arrived on the scene only a few hours earlier—had already discovered (though not yet disclosed) the whereabouts of the champion race horse Silver Blaze, who had disappeared just days before its biggest race. With this typically Sherlockian reply—that the event worth noting was the absence of an event—Conan Doyle lets us know that the great detective had also solved the mysterious death of the horse's trainer. Holmes's paradoxical remark likewise sums up the relationship between the *Belmont Report* and the work of the President's Commission for the Study of Ethical Problems in Medicine and Biomedical and Behavioral Research: the curious aspect of the *Belmont Report* is that it played virtually no part in the deliberations or conclusions of the President's Commission.

This absence of influence seems puzzling for several reasons. First, the *Belmont Report* was completed during the period when Congress was deliberating about the creation of the President's Commission and setting that commission's mandate. Second, two members of the National Commission were among the eleven commissioners originally appointed to the president's commission by President Carter. Third, the literature of bioethics was much less extensive than it is today, so the *Belmont Report*—though perhaps too young to have achieved the iconic status it now seems to enjoy alongside its sister publication, Beauchamp and Childress's *Principles of Biomedical Ethics*, which was published a few months later—was nonetheless one of the first and most prominent attempts to provide a theoretical underpinning for the field. Why, then, was the *Belmont Report* the dog that failed to bark in the night? To answer that question, we need to look at the origins of the President's Commission and the way it undertook to carry out its mandate.

The Origins

The statute creating the President's Commission was approved by the Senate and the House of Representatives on October 15, 1978,[1] just fifteen days after the National Commission had conveyed the *Belmont Report* to the president and the presiding officers of the Senate and House, as well as to the Secretary of Health, Education, and Welfare, in whose department the National Commission resided. The National Commission grew out of a series of hearings held in 1973 by Senator Edward Kennedy's health subcommittee of the Senate Committee on Human Resources.[2] At these hearings, witnesses documented "a wide variety of abuses in the field of human experimentation,"[3] principally involving prisoners, veterans, mental patients, patients at birth control clinics, and participants in the notorious Tuskegee Syphilis Study, whose forty-year ordeal had come to light in 1972. While leading scholars on human subject research, such as Bernard Barber and Jay Katz, and leaders of the research community endorsed what was then called the protection of human subjects bill, the Department of Health, Education, and Welfare (DHEW) opposed it. In order to avoid legislation, the department was then scrambling to develop regulations to ensure greater protection of human subjects than had been achieved under the guidelines it had developed over the preceding decade for biomedical research institutions contracting with the research grants office at the National Institutes of Health (NIH). A compromise was crafted in which the Senate subcommittee held off seeking legislated rules for human subjects research (although it did require that research institutions set up Institutional Review Boards [IRBs]) and instead created a commission that not only would study the protection of research subjects but would possess action-forcing power, under which the secretary of HEW was required to publish the commission's recommendations in the *Federal Register* and then either adopt those recommendations in the form of appropriate regulations or explain why such a step was rejected. When the bill came before Congress in 1974, further impetus for its passage had been added by reports of research using decapitated fetuses. Responding to lurid accounts of these studies, senators argued for a ban on any federal support of fetal research. In a compromise literally worked out on the spot, it was agreed to impose a moratorium on such funding while the National Commission studied the subject and reported back its conclusions and recommendations. In order to ensure that this task did not drag out, the National Commission was given just four months from the time of its appointment to complete this report—a deadline that it succeeded in meeting.

By the beginning of 1978, the National Commission had also completed its mandated reports on research with children and with prisoners and on the use of psychosurgery in the United States and had six more reports nearing completion before it was due to fold its tents in October.[4] Under the 1974 National Research Act, the National Commission was supposed to be succeeded by a National Advisory Council for the Protection of Subjects of Biomedical and Behavioral Research. This standing panel was to be chaired by the secretary of HEW and to consist of seven to fifteen other individuals selected by the secretary from the fields of "medicine, law, ethics, theology, the biological, physical, behavioral and social sciences, philosophy, humanities, health administration, government, and public affairs."[5] While the National Commission's work had been widely applauded for its analysis, and its action-forcing power had proven successful in producing appropriate regulatory response in several areas from the secretary, leaders in both houses of Congress had for several years expressed serious reservations about the wisdom of turning supervision of the protection of human subjects over to a DHEW advisory council. First, continuing revelations of research abuses — especially in studies conducted on unwitting subjects by the Central Intelligence Agency (CIA) and the Department of Defense — led Senator Kennedy's committee to conclude "that the jurisdiction of the National Commission is too limited and that human subjects are at risk in . . . biomedical research programs which are outside the Commission's jurisdiction."[6] Thus, in place of the National Advisory Council, the committee's bill, S. 2579, proposed to establish a President's Commission for the Protection of Human Subjects, modeled on the National Commission but with government-wide jurisdiction. Second, other legislators — principally Representative Paul Rogers, chair of the health subcommittee of the House Committee on Interstate and Foreign Commerce — had concluded that a presidential panel on bioethics ought to go beyond the National Commission's special study on the implications of biomedical advances and tackle a number of specific topics such as informed consent for medical treatment, the definition of death, genetic testing and counseling, medical privacy, and the availability of health care across economic and geographic groups.[7]

Although it retained a primary emphasis on the protection of human subjects in research, the Senate bill was broadened to include "special studies" of the topics incorporated in the House proposal. During the floor debate on the bill on June 26, 1978, several senators objected to certain of these studies, but motions made by Senator Jesse Helms to eliminate the studies on the definition of death and on genetic counseling and testing failed to be adopted. During this debate, supporters of S. 2579 repeatedly cited the National Commission's record of producing balanced and useful reports on particular areas of research and its lack of regulatory power (a feature carried over to the proposed presidential commission) to counter the notion that it would be dangerous to assign the new commission the studies in question,[8] while the opponents cited the National Commission's reports as evidence of the danger of assigning such tasks to a successor body.[9] During these exchanges, the only reference to any ethical principles arose when Senator Helms introduced an amendment regarding the study on genetic testing and counseling to the effect that "such study shall evidence concern for the essential equality of all human beings, born and unborn."[10] Senator Kennedy agreed to the amendment (which was adopted) on the grounds that "it has always been my sense that the [National] Commission itself has in the past indicated a strong concern for the value of human beings and the essential equality of human beings, born and unborn."[11] As amended, the bill was approved by a Senate vote of 68 to 10, and the House then took up the topic in the form of H.R. 13662.[12]

In testifying on August 4, 1978, before Representative Rogers's subcommittee, Senator Kennedy expressed his concern that the presidential commission that would be established under the House bill did not "have a specific mandate to investigate, study, and report on the protection of human subjects of biomedical and behavioral research."[13] In order to avoid taking "a major step backwards" by eliminating "the only independent focus on protecting human research subjects at a time when our clinical research budgets are expanding to break all records in government and the private sector,"[14] Senator Kennedy urged the House to incorporate the relevant provisions of S. 2579. Reminiscent of the events of 1974, DHEW opposed S. 2579 precisely because "the responsibility for the protection of human subjects of federally supported research should remain with accountable Federal officials" rather than with "a Presidential Commission of private citizens devoting a small portion of their time each year to public service."[15]

Once H.R. 13662 was passed by the House on September 19, it was referred to a joint House-Senate conference committee to resolve its remaining differences with the Senate bill.[16] Given the tenor of the earlier hearings and floor debates, it is hardly surprising that the points addressed by the conference committee were entirely formal or procedural. These included adopting the House version of the name (because the commission's mandate went well beyond the "protection of human research subjects" rubric used by the Senate); specifying commissioners' qualifications (following the House version, a ratio of 6 from medicine and research and 5 from other fields, rather than the Senate's reverse ratio of 5 to 6); providing that the departmental representatives would afford liaison with the Commission (House version) rather than being ex-officio advisors (Senate version); deleting a Senate provision that would have allowed congressional committees to request a study and adding authority for the president to require the Commission to conduct additional studies; specifying the scope and timing of the annual report; and replacing a four-year "sunset clause" with a specific sunset date of December 31, 1982.[17]

The conference committee struck a compromise regarding the duties of the President's Commission. At the top of that body's agenda, it placed the specific studies arising from medical care, adding "differences in the availability of health services as determined by the income or residence of the persons receiving the services" to the Senate's list of topics (informed consent to treatment, definition of death, genetic testing and counseling programs, and safeguarding the privacy of research subjects and the confidentiality of patient records). Acceding to Senator Kennedy's concern that the oversight authority of the commission had been too weak in the House bill, the House agreed that the commission should report every second year on the "adequacy and uniformity" of agencies' policies, regulations, and guidelines on the protection of human subjects; in place of the Senate provision authorizing the commission "to review in detail any program of any Federal department or agency to determine whether human research subjects are adequately protected," the conference committee made the commission responsible for evaluating the "adequacy of the implementation" of such rules, policies, and guidelines.

The three issues before the conference committee that were most closely related to the National Commission did not go beyond process to substance. First, the conference committee deleted a Senate provision that would have deemed the remaining members of the National Commission members of the President's Commission on an interim basis, until the president appointed new members. Dr. Kenneth Ryan, the chair of the National Commission, had testified in August that the National Commission was in the process of disbanding and did not want to take on a lame duck role that would interfere with a successful launch of the new body.[18] Instead, the revised bill followed the House version in requiring the president to appoint members within ninety days of the statute's enactment.[19] Second, the conference committee dropped the provision that the President's Commission should complete any duties of the National Commission, since the latter had finished all the reports specified in section 202 of its enabling legislation.[20] Finally, the compromise bill also dropped the Senate's requirement that the President's Commission "undertake studies of the ethical, social and legal implications of advances in biomedical and behavioral research and technology," which mandated the commission to "consider a very broad spectrum of relevant issues." The conference committee concluded that this requirement was "unnecessary" because it would result in the new commission duplicating the work the National Commission had completed on its special study, "the report on which is about to be issued."[21] (This is a reference not to the *Belmont Report* but to another report issued the same day, the so-called special study, which resulted from the assignment that fell furthest outside the core of the National Commission's work, that received the least attention from the commissioners — consisting instead principally of a contracted project carried out jointly by two think tanks, a national opinion poll, and a "scholarly adjunct" prepared by a core group of Harvard and Boston University faculty, based on a colloquium of twenty-five leading scholars and scientists — and that had the least apparent impact.[22])

Two points stand out regarding the legislative process and the assignment of responsibility and authority to the President's Commission. The first is the virtual absence of any mention, much less discussion, of the *Belmont Report* in the legislative process itself. When Senator Kennedy's subcommittee considered S. 2579 and when the bill was debated on the floor, he and other senators evidenced a great deal of familiarity with the work of the National Commission, yet the only reports to which they made reference were those that addressed special populations and the like (such as the psychosurgery report). Likewise, when Dr. Ryan appeared in August 1978 before the House subcommittee, even as the *Belmont Report* was being readied for submission, it went unnoticed. The summary of the National Commission's activities submitted by Dr. Ryan to the House subcommittee is the sole acknowledgment of the report, and that consists of nothing more than the observation that among the four reports "in the final stages of development" (as of August 1, 1978) was "a statement of the basic ethical principles and guidelines for the conduct of biomedical and behavioral research with human subjects (the *Belmont Report*)."[23] The National Commission's chairman did not highlight the *Belmont Report* in his oral statements, and none of the members of Congress chose to inquire about these "basic ethical principles" or their possible relevance to the choices that Congress faced in shaping the legislation before it.

Second, the legislation establishing the President's Commission did not itself include any reference to "principles," much less to the *Belmont Report*. In this respect, the 1978 statute differed from the 1974 law that was supposedly being expanded. The earlier act had, after all, placed at the top of the National Commission's duties "a comprehensive investigation and study to identify the basic ethical principles which should underlie the conduct of biomedical and behavioral research involving human subjects."[24] Ironically, the only mention of bioethical principles in the 1978 legislation was in the details of the "general studies" that the Senate bill had assigned to the President's Commission. In performing its study of the ethical, social, and legal implications of biomedical advances, the commission was required to include "an analysis and evaluation of laws and moral and ethical principles governing the use of technology in medical practice, and of public understanding of and attitudes toward such laws and principles."[25] Yet this was the very language struck by the conference committee — without any recognition that it was stripping from the new commission its only requirement to examine principles — on the grounds that the National Commission's forthcoming "special study" made the provision "unnecessary."

The Work Process of the President's Commission

When the President's Commission began its work, each member received a set of the reports of the National Commission, including the *Belmont Report*. Among the eleven commissioners were two — professors Albert Jonsen and Patricia King — for whom it was familiar reading, as they had served on the National Commission. Yet despite this distribution and the presence of two "authors," the members of the President's Commission apparently did not find the *Belmont Report* particularly useful in guiding their deliberations or analysis. Why was that, and, given the relatively meager alternatives, to what other sources did the commissioners turn for guidance?

One explanation might be that the *Belmont Report* was seen as not being on point for the primary task of the President's Commission, that is, those features that set its responsibilities apart from those of the National Commission, namely, examining the ethical, social, and legal issues in health care rather than in research. After all, the *Belmont Report* states at the outset that it concerns "three principles, or general presumptive judgments, that are relevant to research involving human subjects."[26] The *Belmont Report* then asserts that it "is important to distinguish between biomedical and behavioral research, on the one hand, and the practice of accepted therapy on the other."[27] Although this explanation might have some surface appeal, it seems very dubious to me, for two reasons. First, the report made clear that these principles were drawn from "those generally accepted in our cultural tradition," because they are "particularly relevant to the ethics of research involving human subjects,"[28] not because they are only relevant to research — as indeed there would be no reason to think they would be so limited. Second, the different, albeit superficially rather similar, set of principles in the book by Beauchamp and Childress, which was in print by the time the President's Commission began its work and which was familiar to the staff of the commission as well as to at least a few of its members, made clear the ways in which Belmont-style principles did apply to medical practice as well as to research. Therefore, I do not think we can find the explanation in the conclusion that the *Belmont Report* was simply "not on point" for the work of the President's Commission.

A second explanation is that the commissioners and staff simply did not find the report very useful.[29] I know that is the view of the academic philosophers who served in turn as the first two staff ethicists for the President's Commission. They simply did not rely on the report because, in the words of one, they found the principles endorsed overly broad, vaguely defined, and unranked, "so that the trade-offs that occur must be decided on the basis of 'judgment.'"[30] This view is not the same as a rejection of principles. For example, John Rawls's theory of justice establishes a ranking of principles and requires that each in turn be obeyed before the next principle is even considered, thereby producing a theory that reduces the role of intuition or judgment. Similar observations could be made about utilitarianism and about certain religious systems. Thus, to the extent the staff or members of the President's Commission were inclined to draw on moral theory — at least in the more philosophical parts of our work — the inclination was to draw on general normative theory rather than on the theoretical work in bioethics, of which there was very little at the time.

Furthermore, the President's Commission behaved in much the same way as the body that had authored the *Belmont Report* itself, which was inductive rather than deductive. The National Commission did not begin by agreeing on the three principles and deriving specific guidance for each topic it addressed from an application of the principles, even if a literalist reading of its enabling statute suggests that this may have been the way that Congress thought an ethics commission would behave. Likewise, the work of the President's Commission on its many topics never began by deducing conclusions from general principles, whether of our own identification or derived from the *Belmont Report*.

Finally, and perhaps most important, our method of analysis was not merely inductive but highly contextual. That is, it depended on careful examination of the specifics of the various topics and a search for practical conclusions about them — not abstract philosophy — that made sense in the context of the particular topic. Sometimes that analysis would indeed lead us to moral principles, such as respect for personal self-determination, that appear in the *Belmont Report*. But we discussed such principles and used them to inform our conclusions not because they were found in that report but because of their long and rich history in moral philosophy, which is the same reason that commended them to the National Commission. The basic point, however, is that from the beginning of our work, in the report on the determination of death, the philosophical issues that absorbed the President's Commission were those deeply embedded in the context of the topic itself — for example, what characteristics are important to personhood or human identity, such that their loss marks the transition from life to death of a particular person? Our philosophical analysis was always in the service of practical conclusions and hence had to correspond with analysis based on medical, legal, and social factors as well — not just "informed consent" as an exercise in "respect for persons," but as a doctrine that has emerged from legal cases in which it arose in certain ways, as a social practice in medicine that is experienced in certain ways by patients and healthcare professionals, and so forth.

One can find all three of the *Belmont* principles in use in our reports, but one can also find other themes recurring that play out additional moral tensions: grappling equitably with limited resources, making decisions in the face of uncertainty and different levels of technical expertise, and utilizing public versus private decision-making processes.[31] For all of these reasons, the *Belmont Report* was (along with the special study) probably the least influential with the President's Commission of the ten or so documents produced by the National Commission. Had we been required to state our own "principles" — that is, had the provisions about "the general study" included in S. 2579 survived in the statute as enacted — it is possible that we would have come up with a document not so different from the *Belmont Report*, though perhaps we would have owned up to the fact that our method was case- and situation-based. I believe that our work — like that of the National Commission — is better judged by twin criteria: are the reports individually sensible (i.e., do they persuasively present ethically defensible resolutions for particular topics that can be practically implemented?), and do the reports stand comfortably as a group, deriving some support, one from the next, and avoiding any outright contradictions? While governmental bioethics commissions may be able to draw generalizations from their work and even derive ethical principles, it seems to me that they are not really well suited to crafting philosophical theories as such.

The Mystery Resolved

One alternative explanation for the failure of the President's Commission to rely on the *Belmont Report* deserves mention, if only to be rejected. Knowing that one governmental commission ignored the work of another, a reasonable person might well conclude that the explanation lies in the differing ideologies of their members or simply the different partisan affiliations of the two bodies. Yet such was not the case here. Both the National Commission and the President's Commission were created by acts of Congress championed by the same legislators, most important Senator Kennedy. The two groups of persons appointed were not distinctively different in background or viewpoints; indeed, two of the eleven members of the first commission were members of the second group of eleven commissioners. The commissions were not seen as partisan bodies in the political sense. The National Commission began its work under a Republican president (Nixon), but completed most of that work under his Democratic successor (Carter), whereas the President's Commission was appointed by that Democratic president, yet submitted all of its reports to his Republican successor (Reagan). (The lack of partisanship can also be seen in that all of those reports were approved unanimously — save one, which had one dissenting vote — even though, by the end of the President's Commission's life, only three of the eleven commissioners were among those originally appointed by President Carter.) This continuity extended to the staff: three of the National Commission's professional staff, as well as many of its advisers and formal consultants, played the same roles for the President's Commission.

In sum, the President's Commission was seen, both by those who created it and by those who served on and staffed it, as the successor to the National Commission, albeit with a broader mandate in terms of subject matter and in terms of jurisdiction (being not merely departmental but presidential and hence government-wide). The President's Commission was familiar with all the work of its predecessor and saw part of its mandate as helping to bring that work to fruition, for example, by pressing federal officials to respond to recommendations of the National Commission that remained in limbo. Thus, the easy explanation — that the President's Commission represented some sort of break or sharp turn from the National Commission — simply does not apply.

Given the strong ties between the commissions and the sense of continuity in their work, the failure of the President's Commission to make much if any use of what is now widely regarded as the National Commission's major product does seem like the dog that failed to bark in the night. Yet viewed in the context both of the times (as seen in the scant attention paid to the *Belmont Report* by legislators in framing the tasks of the President's Commission) and of the latter commission's objectives and work processes, this failure is much less puzzling.

Notes

1. Community Mental Health Centers Extension Act of 1978, Public Law 95-622, Title III, 92 Stat. 3437, codified at 42 USC 300v et seq.

2. One aspect of the National Commission's work—the mandate that it conduct what was termed the "special study," a report on the implications of advances in biomedical and behavioral technology—went back to a bill first introduced by Senator Walter Mondale in 1968 that would have set up a body to examine the social, ethical, and legal ramifications of developments in biomedical science and clinical practice. Senator Jacob Javits also championed such a provision and ensured that it was included in the 1974 act and again in 1978 in S. 2579, from which it was deleted in the House-Senate conference, see pp. 12 and 14.

3. Senate Committee on Human Resources, 95th Congress, 2nd sess., President's Commission for the Protection of Human Subjects of Biomedical and Behavioral Research Act of 1978 (Report No. 95-852), May 15, 1978, p. 1.

4. The National Commission's original thirty-six-month lifespan, which had been twice extended by Congress, was due to end on October 31, 1978.

5. Public Health Service Act 217(f)(1). The National Advisory Council was established by Title II, Part A of the National Research Act, which was codified at 42 USC 289l-1.

6. S. Rep. 95-852, at p. 4.

7. President's Commission on Bioethics, Hearing before the Subcommittee on Health and the Environment, House Committee on Interstate and Foreign Commerce, 95th Cong., 2nd sess., on H.R. 13662, August 4, 1978, at 1 (Statement of Rep. Paul G. Rogers) [hereafter House Hearing]:

> While ethical problems connected with research on human subjects still are very much with us, our society is currently grappling with many very difficult ethical dilemmas in the practice of medicine and in the conduct of biomedical and behavioral research which go far beyond the protection of the right of the research subject.

8. See, e.g., 95th Cong., 2nd sess., *Congressional Record* 124 (June 26, 1978): 18897 and 18911-12 (remarks of Sen. Kennedy); 18913 (remarks of Sen. Williams).

9. Ibid., 18907 (remarks of Sen. Helms: "[T]he assertion that the activity of the National Commission . . . is noncontroversial is not entirely accurate," citing its report on fetal research).

10. Ibid., 18906.

11. Ibid., 18907.

12. In the 94th Congress, the Senate had passed a similar bill, but it died when the House failed to complete action on its version.

13. House Hearing, at 37 (testimony of Sen. Edward M. Kennedy).

14. Ibid.

15. Ibid., 31 (testimony of James C. Gaither, Chairman, Ethics Advisory Board [EAB], DHEW). In questioning of Mr. Gaither, Chairman Rogers established that he was an attorney in San Francisco, not a federal official, and hence spent only a portion of his time on the EAB. Mr. Gaither replied that the EAB was not analogous to the proposed presidential commission because the latter was mandated to "monitor" federal research. Ibid., 34. The administration also objected to S. 2579 because it "requires each Cabinet officer to respond independently rather than the administration as a whole, to each recommendation and without any provision for uniformity." Charles McCarthy, who, as executive director-designate of the EAB, accompanied Mr. Gaither to the hearing, effectively contradicted his boss' argument by pointing out that

> the National Commission has recommended that HEW, which supports approximately 40 percent of all research involving human subjects, exercise a lead role among Federal research agencies. Now that leadership would be presumably under the guidance of a commission or some other body. Rather than have the Presidential Commission deal independently with each of the Federal research agencies, it would deal with all of them through the leadership of HEW. (Ibid., 34–35)

16. The provisions of S. 2579 were also incorporated into S. 2450, which was approved the same day, and the conference committee took Title III of the latter bill as the framework for the bill it recommended for adoption by both chambers.

17. Although a Senate bill to amend this date to December 31, 1984, became entangled with other, unrelated legislative business during the fall of 1982, the sunset date was pushed back to March 31, 1983, at the eleventh hour. Public Law 97-377 (December 20, 1982).

18. House Hearing, 22 (testimony of Dr. Ryan) and 25 (August 8, 1978, letter from Dr. Ryan).

19. According to this requirement, commissioners should have been named by January 15, 1979; they were actually appointed more than seven months after that date, and the Senate gave its advice and consent to the appointment of Morris B. Abram as chairman on September 29, 1979. Because of delays in arranging for FY 1980 funds appropriated for other agencies to be reprogrammed to the President's Commission, the members were not formally sworn in until the commission's first meeting on January 14, 1980. During this period, the Office of Management and Budget, which had opposed the creation of the President's Commission on the ground that oversight of human subjects research was being provided by the EAB in DHEW, pressed the department to justify the continued existence of the EAB once the President's Commission had been authorized. Secretary Califano, who appointed the EAB, had resigned in the interim, and his successor, Patricia Roberts Harris, acquiesced and abolished the EAB in 1980. At the President's Commission's first meeting, Dr. David Hamburg, the EAB's vice-

chair, described a topic it had agreed to study but would have to abandon — compensation of injured research subjects — and the President's Commission decided to undertake the study, as the commission's statute authorized it to add topics to those mandated by Congress or assigned it by the president.

20. This appears to be a small overstatement on the part of the conference committee, because the National Commission appears not to have fulfilled the duty established by subsection 202 (a)(3) to "conduct an investigation and study to determine the need for a mechanism to assure that human subjects in biomedical and behavioral research not subject to regulation by the Secretary [of HEW] are protected" and if it "determines that such a mechanism is needed, . . . develop and recommend [it] to Congress." While the need for uniform, government-wide protections was widely acknowledged, the National Commission advanced no concrete proposal for research supported by government departments and agencies outside DHEW, much less to protect subjects in privately funded research.

21. Community Mental Health Centers Extension Act of 1978, Joint House-Senate Summary and Explanation, (P.L. 95-622), 95th Cong., 2nd sess., *Congressional Record* 124 (October 14, 1978): H 13566.

22. National Commission for the Protection of Human Subjects of Biomedical and Behavioral Research, U.S. Department of Health, Education, and Welfare, Special Study: Implications of Advances in Biomedical and Behavioral Research (Washington, DC: U.S. Government Printing Office, September 30, 1978).

23. House Hearing, 13 (statement from the National Commission, dated August 1, 1978).

24. National Research Act of 1974, 202(a)(1)(A)(i).

25. S.2579, 1804(a)(1)(C), 95th Cong., 2nd sess., as passed by the Senate, June 26, 1978.

26. National Commission for the Protection of Human Subjects of Biomedical and Behavioral Research, U.S. Department of Health, Education, and Welfare, *The Belmont Report: Ethical Principles and Guidelines for the Protection of Human Subjects of Research* (Washington, DC: U.S. Government Printing Office, September 30, 1978), 1.

27. Ibid., 2.

28. Ibid, 4.

29. Not satisfied with my own recollection that this was the case, I checked with Daniel Wikler and Dan Brock, who were the President's Commission's first two staff philosophers; they confirmed my view, and I draw heavily on their comments in what follows, though they bear no responsibility for the way I articulate my conclusions.

30. Daniel Wikler, personal communication, March 31, 1999.

31. President's Commission for the Study of Ethical Problems in Medicine and Biomedical and Behavioral Research, *Summing Up: The Ethical and Legal Problems in Medicine and Biomedical and Behavioral Research* (Washington, DC: U.S. Government Printing Office, 1983), 71–81.

4 Beyond *Belmont*

Trust, Openness, and the Work of the Advisory Committee on Human Radiation Experiments

Ruth R. Faden
Anna C. Mastroianni
Jeffrey P. Kahn

By the time the Advisory Committee on Human Radiation Experiments (hereafter Advisory Committee) was convened by President Clinton in January 1994, the *Belmont Report* had assumed almost constitutional status internationally as the dominant framework for evaluating the ethics of research with human subjects.[1]

The purpose of this chapter is to interpret the work of the Advisory Committee against the legacy of the *Belmont Report*. We begin by briefly describing the mission and context of the Advisory Committee, as these frame how the *Belmont Report* related to our work. We then turn to three particular aspects of our efforts: the ethical framework adopted by the Advisory Committee, our historical analysis of research involving children, and our analysis of the current state of human subjects research. We conclude with a consideration of two critically important moral themes — openness and trust — that emerged from both the historical and the contemporary work of the Advisory Committee and that are not reflected in the *Belmont* framework.

Mission and Context

Like the National Commission for the Protection of Human Subjects of Biomedical and Behavioral Research (hereafter National Commission), which authored the *Belmont Report*, the Advisory Committee was created in response to allegations of human subjects abuses. In our case, the allegations focused on radiation research conducted or sponsored by the government during the cold war, raising the specter of secret government experiments linked to national security interests.[2] The suspect research included a series of experiments in which hospitalized patients had been administered plutonium and other atomic bomb materials, nontherapeutic research with prisoners, and total body irradiation research with cancer patients.[3]

President Clinton directed the Advisory Committee to uncover the history of these and other human radiation experiments during the period 1944–1974. It was in 1944 that the first government-sponsored human radiation experiments of interest were planned (the plutonium experiments) and in 1974 that the Department of Health, Education, and Welfare adopted regulations governing the conduct of human research, a watershed event in the history of federal protections for human subjects. In addition to asking the Advisory Committee to investigate human radiation experiments, the president directed us to examine cases in which the government had intentionally released radiation into the environment for research purposes. These "intentional releases" also occurred in the period from 1944 to 1974.

The president further charged us with identifying the ethical and scientific standards for evaluating these events and with making recommendations to ensure that whatever wrongdoing may have occurred in the past could not be repeated. This was yet another similarity between the Advisory Committee and the National Commission. The National Commission was required by Congress "to identify the basic ethical principles which should underlie the conduct of biomedical and behavioral research involving human subjects," whereas we were required by executive order to identify ethical "standards."[4]

The *Belmont Report* did not figure formally in our evaluation of the ethics of the historical radiation experiments. In response to the executive order, the Advisory Committee developed its own framework for ethical judgment, a framework that arguably shares a common moral tradition with the *Belmont Report*. Not surprisingly, however, the influence of the *Belmont Report* is evident in aspects of our historical evaluation and is prominent in our evaluation of the current state of protections for human subjects.

The Advisory Committee's Ethical Framework and the *Belmont Report*

Developing an ethical framework for judging the human radiation experiments proved to be one of the Advisory Committee's most difficult tasks. Unlike the *Belmont Report*, our context required engaging controversial issues in moral theory about relativism and the legitimacy of moral judgment across time. We had to address the thorny questions of whether and how to assess the ethics of events (and the people responsible for them) that occurred in what was arguably a different moral era with respect to research involving human subjects, the era before the *Belmont Report*.

The Advisory Committee's response to this task was to adopt an ethical framework for judgment that, like the National Commission's *Belmont Report*, featured basic ethical principles. However, unlike the *Belmont Report*, the Advisory Committee's ethical framework also included two sources of relevant standards in addition to basic principles, or three moral standards overall:

1. Basic ethical principles that are widely accepted and generally regarded as so fundamental as to be applicable to the past as well as the present;

2. The policies of government departments and agencies at the time; and

3. Rules of professional ethics that were widely accepted at the time.

As a specific response to a particular charge in the National Commission's enabling legislation, the commission's focus on ethical principles (and the relationship of these principles to the practices of consent, risk-benefit analysis, and the selection of subjects) is understandable. The National Commission was asked to identify relevant ethical principles, and in the *Belmont Report* the commission delivered on this obligation. The Advisory Committee, by contrast, was asked to identify relevant ethical standards for making moral judgments about events that had occurred as many as fifty years earlier. Our executive order did not presume that moral principles had any necessary role to play in setting such standards; this was a determination left to the Advisory Committee.

Much as the National Commission, we assumed, without justification or defense, the importance of basic principles to moral life. We interpreted these basic ethical principles as general standards or rules that all morally serious individuals accept. We asserted in our final report that the principles we engaged are basic because any minimally acceptable ethical standpoint must include them.

The Advisory Committee identified six basic ethical principles that comprised Standard 1 of our ethical framework:

1. One ought not to treat people as mere means to the ends of others.

2. One ought not to deceive others.

3. One ought not to inflict harm or risk of harm.

4. One ought to promote welfare and prevent harm.

5. One ought to treat people fairly and with equal respect.

6. One ought to respect the self-determination of others.[5]

Two criteria were implicitly used to select these six principles. First, as our specific context was retrospective moral judgment, it was important to us to identify principles that even a moral relativist would agree were pervasive features of the moral life of the United States fifty years ago, just as they are today. Second, we were concerned to identify from among such enduring moral principles those moral commitments of particular relevance to our work. It was not our view that these six represented the only moral principles worthy of consideration, or even the most significant. They were, however, the principles that in our view had direct bearing on the moral questions at the heart of the events we were asked to evaluate. Indeed, these principles reflected the moral values that emerged and that the Advisory Committee was actually using as it grappled with specific case studies. As a chronological matter, the Advisory Committee was well into its analyses of these cases by the time the list of principles was formally articulated.

Those of us who did the initial drafting of this part of the Advisory Committee's report did not work with the *Belmont Report* as a model. However, it is possible to regroup the six principles identified by the Advisory Committee so as to make them conform to the three principles of the *Belmont Report*. For example, some might consider the principles "One ought not to treat people as mere means to the ends of others," "One ought not to deceive others," and "One ought to respect the self-determination of others" as all falling under the *Belmont Report* principle of "respect for persons." Such a structure was never considered and doubtless would have been rejected by the Advisory Committee as ill suited to our context as well as analytically unhelpful. How basic moral injunctions relate to one other, whether some principles can be seen as being justified by still "more basic" formulations, or alternatively whether such principles are conceptually or morally distinct remain unresolved in moral theory. In many respects, and certainly in practical contexts, the way basic principles are identified and framed relates more to the purposes they are intended to serve than to any underlying moral structure of justification or theory. In the work of the Advisory Committee it was, for example, critical that we highlight the moral significance of deception, as deception figured prominently in the allegations of abuse we were charged to evaluate. Moreover, differentiating the principle "One ought not to deceive others" from the principles "One ought not to treat people as mere means to the ends of others," and "One ought to respect the self-determination of others" added considerable power to our moral analysis by allowing us to feature independently these different sources of moral transgression.

Standards no. 2 and no. 3 of the Advisory Committee's ethical framework represented our commitment to the view that moral judgment must be contextualized to be meaningful. The Advisory Committee's methodology relied heavily on the integration of moral analysis with rigorous historical analysis and, where possible, empirical scholarship as well.

The Advisory Committee's Historical Analysis of Research Involving Children

The Advisory Committee developed its methodology in the context of specific case studies, each chosen for the particular issues they raised, the research populations they included, and the types of radiation exposure they involved. As an illustration, we consider here one such case, the Advisory Committee's analysis of research involving children. The Advisory Committee's approach to research on children — a group also addressed by the National Commission — offers not only an example of how the Advisory Committee carried out its ethical analysis but also a case for comparison of how the Advisory Committee's approach compares to the *Belmont* approach.

The Advisory Committee chose to focus on children as research subjects for several reasons, two of which are instructive for how the application might compare with that of the *Belmont* framework: children were more likely to be used in research as a means to the ends of others due to their relative dependency and powerlessness (raising issues of adequate consent, selection of subjects, and fair treatment), and the risks posed by radiation research were increased for children (raising issues of acceptable risk and benefit).[6]

The Advisory Committee limited its historical review of radiation research carried out on children to only research without potential direct medical benefit to the subjects — what was traditionally termed nontherapeutic research. This limitation made sense in the context of the Advisory Committee's charge, and because, as a category, such research raised the greatest ethical concern. To do this, twenty-one nontherapeutic experiments were selected from eighty-one radiation research projects carried out on children identified by reviews of government documents and the medical literature from the period of the Advisory Committee's research. Of the eighty-one identified, thirty-seven were considered nontherapeutic, of which twenty-one were funded or carried out by the federal government and so were within the Advisory Committee's charge. Two of these remaining twenty-one were studies conducted at the Fernald and Wrentham schools, about which more information is known than in the other cases due to research carried out by the Massachusetts Task Force on Human Subject Research. A detailed account of those cases can be found in the task force's report.

In evaluating the ethics of these twenty-one experiments, the Advisory Committee relied on the three standards described earlier in this chapter: whether particular research adhered to basic ethical principles that are widely accepted and generally regarded as so fundamental as to be applicable to the past as well as the present (standard no. 1), whether particular research met the policies of government departments and agencies at the time (standard no. 2), and whether particular research met the rules and standards of professional ethics that were widely accepted at the time (standard no. 3). Each of these will be briefly addressed in the context of research on children.

STANDARDS OF PROFESSIONAL ETHICS FOR RESEARCH ON CHILDREN PRIOR TO 1974

In the post–World War II period, biomedical research — including research on children — grew rapidly. But standards regarding the treatment of research subjects were slower to develop. Practice seemed to assume that researchers did not need to obtain the permission of patients (or their parents in the case of children) before using them as research subjects. This was true of both therapeutic and nontherapeutic research, as it was traditionally distinguished. One pediatric researcher from the 1950s who was interviewed by the Advisory Committee observed that many pediatricians believed that permission of parents was not necessary before using a pediatric patient as a research subject, justified in part by a "broader 'ethos of the time' in which 'everyone was a draftee' in a national war on disease."[7]

Such a view appears not to have been universally shared, however, as can be seen in the record of a conference hosted in 1961 by Boston University's Law-Medicine Research Institute (LMRI) on the topic of "Social Responsibility in Pediatric Research." Some pediatric researchers reported regularly obtaining parental permission, and one attendee at the conference admitted to the group that his failure to get permission for lumbar punctures on newborns was wrong and motivated by the concern that there would be no volunteers for such research.[8] If the importance of parental consent was not universally held, the need to reduce the risk to which children were exposed in research seems to have been a widely shared value, at least by those at the LMRI conference. This was consistent with what appeared to be the only early written guidelines regarding children in research, established in the late 1940s by the Atomic Energy Commission (AEC).

EARLY POLICIES OF GOVERNMENT DEPARTMENTS AND AGENCIES FOR RESEARCH ON CHILDREN

In 1949, the AEC instituted rules for evaluating the proposed use of radioisotopes in research on "normal" children. These rules discouraged the general use of radioisotopes in normal children, but suggested that the AEC committee charged with approving use of radioisotopes in research would consider proposals in which (1) the research question could not be studied in other ways and (2) the radiation level would be low enough to be considered "harmless." Note that there was no mention of consent to research in the rules, which instead focused its protections on minimizing risk to subjects.

ASSESSING RADIATION RESEARCH ON CHILDREN USING THE SIX ETHICAL PRINCIPLES IDENTIFIED BY THE ADVISORY COMMITTEE

Violations of several of the Advisory Committee's ethical principles emerged in the committee's analysis of three key components of research involving human subjects: risk, authorization, and selection of subjects. Because of the lack of detailed information in all but one case, only general conclusions could be drawn about the twenty-one experiments as a group. The exception was in the case of the Fernald School where, because more information was available, the Advisory Committee was able to draw more specific conclusions.

With respect to risk, little was known at the time these twenty-one experiments were being conducted about the effects of radiation exposure on children, especially related to what we now know are its longer term health effects, such as excess cancer risk. The Advisory Committee concluded that "investigators had a limited understanding of the potential long-term risks of low-dose radiation and of methods to accurately calculate tissue doses in children."[9] This limited understanding is what motivated researchers to carry out radiation research on children, because radiation effects were the point of at least some of these studies.

Moreover, even burgeoning understandings of radiation risk, coupled with a sense that children needed special protections as embodied in the AEC policy regarding use of radioisotopes in children, argue that researchers had an increasing sense of the risk their research posed to subjects. If this was — or at least should have been — the case, then researchers should have been more cautious in their use of radiation in research on children. This lack of caution implicates two more of the principles identified by the Advisory Committee: "One ought not to inflict harm or risk of harm" and "One ought to promote welfare and prevent harm."

With respect to the question of authorization, the only research project in which there existed sufficient information to draw conclusions involved the Fernald School, a state-operated institution for boys (at least some of whom had mental impairments) outside Boston. Studies at Fernald involved researchers from the Massachusetts Institute of Technology (MIT) working with Fernald staff to carry out nontherapeutic studies of nutritional metabolism. Between 1946 and 1953, seventy-four students were exposed to radioactive iron or calcium in breakfast cereal they ate as members of the Fernald "science club." While it is clear that the doses of radiation involved were — and so risk was — low, the evidence regarding parental consent suggests that the information provided to parents about the research was seriously deficient.

The Massachusetts Task Force did extensive research into the history of the Fernald research and found two letters to parents seeking permission for the research participation of their children. The letters indicate that the children would receive a special diet rich in various minerals and vitamins, which would require periodic blood samples "similar to those to which our patients are already accustomed, and which will cause no discomfort or change in their physical condition other than possible improvement."[10] There is no mention of exposure to radiation or of any other risks of the research in either letter, and the implication in both is that there will be some benefit to the children.

The Massachusetts Task Force concluded, and the Advisory Committee agreed, that there was insufficient information about the research for the parents to make an informed decision and that the researchers failed to adequately inform parents that the research was nontherapeutic. Because the consent process could easily have led parents to believe that research participation would benefit their children, it bordered on violation of the committee-identified principle "One ought not to deceive others." The lack of adequate consent meant that parents, as proxy decision makers for their children, were unable to make informed decisions, so whatever authorization they gave would not meet the principle requiring that "One ought to respect the self-determination of others." Perhaps most powerfully, the Advisory Committee concluded that this research violated the long-held commitment in biomedical research that it is wrong to use research subjects as the mere means to the ends of others. In no respect could this research be construed as serving the interests or ends of the participating children. Whether these children, or any child, could have validly consented to this use of their persons is historically beside the point. Both the children and their parents were misled about the nature of the research, which made any prospect of legitimate permission impossible.

With respect to selection of subjects, the Advisory Committee was unable to assess the characteristics and recruitment methods of the subject populations making up nineteen of the twenty-one studies examined. However, in the case of research at both the Fernald and Wrentham schools, it is apparent that institutionalized children were selected because they were in environments in which the experimental variables and research subjects could be more easily recruited and controlled. An irony, however, is that while there were and are numerous other boarding schools in the Boston area, boys from Choate and other "elite" schools were not included. The Advisory Committee concluded that

> perhaps the investigators, who were not responsible for the poor conditions at Fernald, believed that the opportunities provided to the members of the Science Club brightened the lives of these children, if only briefly. Reasoning of this sort, however, can only too easily lead to unjustifiable disregard of the equal worth of all people and to unfair treatment.[11]

Thus, the Advisory Committee found that using children in this way violated the principle that "One ought to treat people fairly and with equal respect."

The case of Fernald, in addition to the accepted practice during this era of using hospitalized children as subjects out of convenience, underscore the importance of adequate protections for institutionalized children used in research. The fact that no such special protections exist even today represents a policy void that the Advisory Committee strongly urged the federal government to address.

Even in areas of the research process where policies for the protection of subjects have been promulgated, it is not clear that they adequately protect the rights and interests of research subjects, as we discuss in the following section.

The Advisory Committee's Contemporary Analysis

In addition to historical review and recommendations, President Clinton charged the Advisory Committee with making recommendations to ensure that the wrongdoing that may have occurred in the past could not be repeated. The federal regulations and the conceptual framework based on the *Belmont Report* became so widely adopted that it could be argued that their establishment marked the end of serious shortcomings in federal research ethics policies and practices. This was not, however, the conclusion we drew based on our evaluation of contemporary practices and policies.

We approached our evaluation of the contemporary research world from three different perspectives. At the funding level, we analyzed the research protection policies and practices of the sixteen federal agencies and departments that oversee human subjects research. At the local review and oversight level, we examined a sample of research proposals submitted to and approved by Institutional Review Boards (IRBs) and the corresponding IRB-approved consent forms. A focus of this review was to understand how well research proposals addressed central ethical considerations such as risk, voluntariness, and subject selection and whether informed consent procedures seemed to be appropriate. To obtain the perspective of former, current, and prospective research subjects, we interviewed nearly nineteen hundred patients receiving outpatient medical care about their experiences with, understanding of, and attitudes about medical research. Our analysis was supplemented by requests to IRB chairs for their perspectives on the implementation of the federally mandated system of oversight.

The Advisory Committee found, not surprisingly, that there have been significant improvements in protections for human subjects in recent decades. The *Belmont* framework and the regulations and the system of oversight that reflect it have provided assurances that subjects' basic rights and interests are respected. But the Advisory Committee's investigation of contemporary research indicated that, while the system largely protects these rights and interests, serious deficiencies remain. Today, many of these concerns arise in the context of research where patients are subjects. Among our many findings and recommendations, two areas in particular deserve further explication: the limitations of informed consent (including understandings of risk and benefit) and the confusion between research and therapy, issues that are highlighted in the *Belmont Report*.

The Advisory Committee found evidence of numerous problems with the informed-consent process. The informed-consent forms we reviewed as part of our research proposal review project were often difficult to understand. Problematic forms were written at inappropriately high reading levels and contained technical language and overwhelming detail. In 14 percent of the protocols we reviewed, and in 30 percent of those that carried more than minimal risk to subjects, the consent forms overpromised the benefits of research and underestimated its risks.

The *Belmont Report* highlighted the importance of distinguishing the boundaries of research and therapy for the threshold purpose of ensuring that research undergo proper review. The Advisory Committee found, however, that even with IRB review, patient-subjects still may not understand the difference between research participation and clinical care and that other factors that were not discussed in the *Belmont Report* may affect this understanding.

In our surveys and interviews of patients, our review of research documents, and in public testimony, we found evidence of confusion over the distinction between research and therapy. For example, the Advisory Committee's survey of nearly two thousand patients showed that of those who were asked whether they were currently or recently in a research protocol, 6 percent reported that they were research subjects when their medical records indicated that they were in fact not part of research. More worrisome were results that another 6 percent of the patients surveyed reported that they were not research subjects when their medical records indicated that they were currently participating in a research protocol. Together, these results point out that nearly one in eight patients in the sample wrongly understood their research status. The Advisory Committee found that this was an unacceptably high proportion, given that patients who consent to research are in real ways agreeing to accept risk so that the information gained can be used to benefit future patients; only on occasion will some of these benefits fall to the subjects themselves.

A failure by patient-subjects to understand the difference between research and therapy can easily lead to false expectations, dashed hopes, and a lack of trust in biomedical research. Now more than ever, this distinction is complicated by the facts that medical care is often part of research participation, for instance, in cancer clinical trials, and that there may actually be some therapeutic benefit realized by research subjects. But even when there is a reasonable expectation of personal medical benefit from research participation, patient-subjects must have a meaningful appreciation of the difference between standard medical care for their health problems and whatever research opportunities are offered. Any confusion between the two can lead to a mistaken sense that research participation necessarily carries the same levels of therapeutic success as standard therapy.

This blurring of research and therapy is also influenced by the fact that researchers sometimes wear two hats in their relationship with patient-subjects: that of the investigator and that of the treating physician. There are numerous examples, both past and present, of ethically questionable research that was able to go forward only because the researchers were physicians caring for patients who also became their research subjects. More recent examples involve psychiatric patients who were entered into placebo-controlled drug trials by their physicians, who were also researchers. It is very difficult for patients to understand how the motivations, incentives, and potential biases of investigators are different from those of treating physicians. And when the same person, wearing the same white coat, who has been caring for a patient suggests that research participation might be a good idea, it is understandable that patients have difficulty distinguishing advice about medical care from requests to participate in research.

Moral Lessons Beyond *Belmont*: Trust and Openness

Two sets of moral considerations defined the experience of the Advisory Committee but are not captured or accounted for in the framework provided by the *Belmont Report* or in the discussions of the National Commission: trust and openness. The Advisory Committee's creation was stimulated by revelations that investigators had acted unethically during the cold war period and that the U.S. government had sponsored research that was clearly deceptive and in some cases officially secret. Like other independent commissions established by the government, a driving force for the Advisory Committee's work was the restoration of public trust, not only in the government but also in the research enterprise.

The critical role of trust is further underscored in the Advisory Committee's empirical research into the contemporary world. Research subjects, particularly those who are patients, assume that a process of oversight is in place to protect them from harm and exploitation. IRBs and investigators that overemphasize autonomy and informed consent forms without attention to the process of consent and the subjects' understanding of risks, benefits, and the distinctions between research and therapy run the risk of undermining this trust. Included among the Advisory Committee's recommendations are the development of mechanisms in our current system of oversight to improve these and other aspects of communication to subjects. Also included are recommendations designed to address deficiencies in the current system, a byproduct of which is the sustainability of trust in the research enterprise. These proposals include changes in IRBs; the interpretation of ethics rules and policies; the conduct of research involving military personnel as subjects; oversight, accountability, and sanctions for ethics violations; and compensation for research injuries. In discussing the particular need for investigators to be guided by and sensitive to ethics, the Advisory Committee states, "At stake is not only the well-being of future subjects, but also, at least in part, the future of biomedical science. To the extent that that future depends on public support, it requires the public's trust."[12]

The theme of trust is emphasized throughout the Advisory Committee's report and is also highlighted by Kass et al. in the ongoing analysis of data from the Advisory Committee's interviews with patient-subjects.[13] As Kass et al. emphasize, trust is fragile and easily lost. It requires attention at all levels, particularly at the investigator level where there is direct contact with the subject. As stated by the Advisory Committee, "There can be no better guarantor of that trust than the ethics of the research community."[14]

The history of U.S. government-sponsored radiation experiments teaches us that regulations and new processes for oversight or improved professional ethics will not prevent ethical lapses or exploitation. The corpus of research ethics needs to have a place for considerations of secrecy and openness. Among the Advisory Committee's recommendations are those that ask for new policies for classified research, including the creation of an independent panel of nongovernmental experts and citizen representatives to review and approve all classified research with human subjects.

History has taught us that even where decisions are not officially classified as secret, where they are made behind closed doors by so-called experts without informed participation by those affected, the absence of openness can be as damaging to public trust as an official secret:

Of at least equal import is the development of a more common understanding *among the public* of research involving human subjects, its purposes, and its limitations. Furthermore, if the conduct of the government and of the professional community is to be improved, that conduct
must be available for scrutiny by the American people so that they can
make more informed decisions about the protection and promotion of
their own health and that of the members of their family. . . . Some of
what is regrettable about the past happened, at least in part, because we
as citizens let it happen. Let the lessons of history remind us all that the
best safeguard for the future is an informed and active citizenry.[15]

Conclusion

One could question whether the deficiencies identified by the Advisory
Committee in the contemporary research protections are related to the value
of principles generally or the implementation and understanding of the
Belmont framework in particular. In the course of its work, the Advisory
Committee had occasion to consider frequently the relationship of regulations and codes to ethical conduct. Some of our most engaging historical
findings involved the uncovering of relatively extensive official statements
and guidelines governing human subjects, statements that had little apparent
effect on research practices. From these observations, the Advisory Committee developed a strong skepticism about regulations and rules. At the same
time, however, the Advisory Committee retained a healthy regard for moral
principles. In stressing the importance of moral principles, the Advisory
Committee was not speaking to contemporary debates about the role of principles in moral theory, method, or analysis. Rather, we were speaking to the
importance of moral principles as public *expressions of moral commitment*,
expressions that bind the nation and the professions in a common moral
vision for the enterprise of human research.

There are only three references to the *Belmont Report* in the Advisory
Committee's final document of more than 900 pages.[16] One of these references is especially significant, however. It is among the Advisory Committee's
core recommendations (Recommendation 9) for protecting the rights and
interests of human subjects in the future and represents the Advisory Committee's view about the enduring value of statements of principle and its
respect for the *Belmont Report*, in particular:

> *The Advisory Committee recommends . . . that efforts be undertaken on a*
> *national scale to ensure the centrality of ethics in the conduct of scientists whose*
> *research involves human subjects.* A national understanding of the ethical
> principles underlying research and agreement about their importance is
> essential to the research enterprise and the advancement of the health of
> the nation. The historical record makes clear that the rights and inter-
> ests of human subjects cannot be protected if researchers fail to appre-

ciate sufficiently the moral aspects of human subject research. . . . It is not clear to the Advisory Committee that scientists whose research involves human subjects are any more familiar with the Belmont Report today than their colleagues were with the Nuremberg Code forty years ago. (Recommendation 9)[17]

Notes

1. The Report of the Advisory Committee on Human Radiation Experiments can be found on the Web at http://tis.eh.doe.gov/ohre/roadmap/achre/index.html. It is available in print as Advisory Committee on Human Radiation Experiments, *The Human Radiation Experiments: Final Report of the President's Advisory Committee* (New York: Oxford University Press, 1996). Subsequent references will be to the published book.

2. Executive Order no. 12891, January 15, 1994, 3 C.F.R. 847 (1994).

3. The Advisory Committee also investigated nontherapeutic research with children, human experimentation conducted in conjunction with nuclear weapons testing, intentional environmental releases of radiation, and observational research involving uranium miners and residents of the Marshall Islands.

4. Public Law 93-348, *U.S. Statutes at Large 88* (1974): 342.

5. Advisory Committee, *Human Radiation Experiments*, chap. 4, esp. 114–15.

6. Ibid., 136–37.

7. Ibid., 201, quoting William Silverman.

8. Ibid., 202, notes 41 and 42.

9. Ibid., 209.

10. Task Force on Human Subject Research, "A Report on the Use of Radioactive Materials in Human Subject Research That Involved Residents of State-Operated Facilities within the Commonwealth of Massachusetts from 1943 to 1973," Document 19, Appendix B. Task Force on Human Subject Research, Commonwealth of Massachusetts Department of Mental Retardation, A Report on the Use of Radioactive Materials in Human Subject Research That Involved Residents of State-Operated Facilities within the Commonwealth of Massachusetts from 1943–1973 (April 1994).

11. Advisory Committee, *Human Radiation Experiments*, 212.

12. Ibid., 495.

13. Nancy E. Kass, Jeremy Sugarman, Ruth Faden, and Monica Schoch-Spana, "Trust: The Fragile Foundation of Contemporary Biomedical Research," special issue, *Hastings Center Report* 26S (September–October 1996): 25–29.

14. Advisory Committee, *Human Radiation Experiments*, 495.

15. Ibid.

16. For more discussion of the findings and recommendations of the Advisory Committee on Human Radiation Experiments, see "Trusting Science: Nuremberg and the Human Radiation Experiments," special issue, *Hastings Center Report* 26S (September–October 1996): 5–10; "Advisory Committee on Human Radiation Experiments," guest editors: Anna C. Mastroianni and Jeffrey P. Kahn, special

issue, *Kennedy Institute of Ethics Journal* 6, no. 3 (September 1996): ix–xi; "The Advisory Committee on Human Radiation Experiments. Research Ethics and the Medical Profession: Report of the Advisory Committee on Human Radiation Experiments," *Journal of the American Medical Association* 276, no. 5 (August 1996): 403–409; Allen Buchanan, "Judging the Past: The Case of the Human Radiation Experiments," *Hastings Center Report* 26, no. 3 (May–June 1996): 25–30; Ruth R. Faden, Susan Lederer, and Jonathan D. Moreno, "U.S. Medical Researchers, the Nuremberg Doctors Trial, and the Nuremberg Code: A Review of Findings of the Advisory Committee on Human Radiation Experiments," *Journal of the American Medical Association* 276, no. 20 (November 1996): 1667–71; Anna Mastroianni and Jeffrey Kahn, "Remedies for Human Subjects of Cold War Research: Recommendations of the Advisory Committee," *Journal of Law, Medicine and Ethics* 24, no. 2 (summer 1996): 118–26.

17. Advisory Committee, *Human Radiation Experiments*, 522–23.

5 Relating to History

The Influence of the National Commission and Its *Belmont Report* on the National Bioethics Advisory Commission

Harold T. Shapiro
Eric M. Meslin

The National Bioethics Advisory Commission (NBAC), 1996–2001, was one in a series of commissions and panels sporadically convened by the U.S. government over the last three decades to provide advice and make recommendations regarding bioethics issues arising from developments in biomedical science, the evolution of clinical care practices, and changing moral sensibilities.[1] NBAC owed a great debt to previous commissions and panels, to a steadily accumulating literature in bioethics, and particularly to such milestone documents as the Nuremberg Code, the several versions of the Declaration of Helsinki, and, especially in this country, the National Commission's *Belmont Report*. In addition, it owed a special debt of gratitude to the work of the National Commission, the President's Commission, the Ethics Advisory Board, and the Advisory Committee on Human Radiation Experiments — the national committees that predated NBAC and whose legacies NBAC partly inherited — and to the generations of scholars who, collectively, greatly expanded NBAC's understanding in this arena. Because NBAC's charter and executive order gave priority to issues arising from research involving human subjects, NBAC naturally shared a special affinity with the work of the National Commission, in particular.

In this chapter we attempt to put into perspective NBAC's contribution to this ongoing legacy. First, we describe how NBAC's establishment fits into the U.S. model for public policy deliberation in bioethics, as initially developed by the National Commission. Second, we discuss how the *Belmont Report* influenced NBAC's reports and recommendations to various degrees. Finally, we will suggest some lessons learned from NBAC's experience.

The establishment of the National Commission and the publication of its *Belmont Report* were primarily a response to concerns regarding the identification and protection of persons serving as subjects in biomedical and behavioral research. There is little question that both the National Commission and the *Belmont Report* have had a significant and ongoing influence on public policies, professional practices, and regulations in this arena, both here and abroad. For the first time in this country, a deliberate confluence occurred among serious ethical scholarship, empirical study, and legal and policy analysis that provided an intellectual and pragmatic approach to a defined set of issues in bioethics. Sometimes lost in current discussions about the adequacy of the *Belmont* principles or the sufficiency of the guidelines developed by the National Commission is an appreciation of how challenging a task it is to secure agreement on profoundly important moral problems in biomedicine and to describe, explain, and justify conclusions in language accessible to a large variety of constituents. Perhaps only the commissioners and staff of those groups that have followed in the National Commission's wake can fully appreciate the value of this pioneering work.

At the same time, the National Commission's work built on practices already established in certain settings and on the prior and ongoing work of scholars in developing approaches to identifying and resolving moral problems in biomedicine that might inform public policy. The National Commission did not invent bioethics, nor did it discover the principles of bioethics often associated with its work. This is not in any way to deny the substantial and lasting contributions of the *Belmont Report* itself, but to acknowledge both the important work that preceded it and the work that was going on at the same time.

The *Belmont Report* represents a special and innovative approach to the use of moral philosophy (i.e., approaches for deciding how we should act and why), not simply regarding private acts that take others' interests into account, but for the rather different purpose of setting public policies that affect the lives of all citizens. After all, it is only through national legislation and/or regulation that all citizens achieve truly equal moral and legal standing. The *Belmont Report* represents a combination of consequentialist approaches (judging prospective public policies by their consequences) and more deontological approaches (judging particular acts, whether public or private, on their own, quite apart from any calculus regarding the battery of outcomes). That is, the *Belmont Report*, and the regulations and public policies that followed from it, represented an attempt to balance several types of ethical considerations.

But in considering the influence of the *Belmont Report*, it is useful to distinguish between its direct influence on policies regarding the protection of human subjects and its more indirect influence through the particular approach it took both to bioethical problems and to public discourse and policy formation on these problems. Clearly, the *Belmont Report* had both an educational function and a role in expanding the public discourse on the intersection of ethical principles and other considerations that also inform public policy. The evidence of this influence is seen in the education and training programs that include questions on *Belmont* required of National

Institutes of Health (NIH)–funded investigators at U.S. institutions, in the assurances negotiated between universities and the federal government that often include *Belmont* as an appendix, and in the required readings in university courses on biomedical ethics.

The *Belmont Report* influenced NBAC's thinking on human subject protection and in the way NBAC tried to provide a comprehensible basis for the ongoing public discussions and public policy decisions on the particular matters we addressed. With respect to particulars, while the *Belmont Report* had its most noticeable influence on NBAC's reports that dealt directly with research involving human subjects, it also had a significant but quite different impact on reports on cloning and on stem cell research.

NBAC was established by executive order, signed by President Clinton in October 1995.[2] Its responsibilities were to advise the National Science and Technology Council — which is chaired by the president — and other government agencies on the appropriateness of governmental policies and regulations relating to bioethical issues arising from research on human biology and behavior, as well as the applications, including clinical applications, of that research. It consisted of eighteen members from a variety of disciplines and areas of expertise appointed by President Clinton.

As a federal advisory committee, NBAC was subject to the Federal Advisory Committee Act (FACA),[3] the federal law established during the Nixon administration that provides the public with an assurance that groups established by the government to advise the government are publicly accountable. The implementing requirements of FACA are designed to prevent special-interest control of the bodies providing advice to the president and executive agencies,[4] as well as to allow public scrutiny of the deliberations of the bodies formulating the advice.[5] Beyond the right to attend advisory committee meetings, members of the public have the right to file statements with advisory committees and to speak before advisory committees.[6] Advisory committees may only make recommendations, as opposed to determining policy, which is reserved solely for "the President or an officer of the Federal Government."[7]

Under FACA's guidelines, all of NBAC's meetings had to be held in public, under not only the watchful eye of the media but also the scrutiny of various groups — from those with particular interests in a topic on our agenda to the staff from government agencies, congressional offices, and health and research organizations. At times this created difficulties, as some commission members found it difficult to explore new and speculative ideas in this environment because of the substantial risk that the media or others would distort their ideas and, perhaps, the work of the commission. Moreover, some interest groups demonstrated that they were not beyond attempts at public intimidation. As a result, there was some confusion on occasion, even among commissioners, about how to balance the duty to inform the public and the duty to proceed with important substantive tasks. However, complete public access to NBAC's deliberations is mainly positive for the following three reasons: (1) public access to deliberations on morally contested issues may increase public confidence in those deliberations and resulting conclusions, (2) public access to discussions enables interested parties to provide more helpful testimony, and (3) public access imposes a certain useful discipline that keeps discussion more focused than may otherwise be the case — a process that served NBAC well, for example, in developing its report on *Research Involving Persons with Mental Disorders That May Affect Decision-making Capacity* [the "Capacity Report"].[8]

NBAC advised the president (through the National Science and Technology Council) and other government agencies and departments on issues arising from research involving human subjects and from the management of genetic information. This advice was provided in the form of recommendations on various issues, the depth and breadth of which depended on the issues and the type of advice requested. For example, the executive order that established NBAC instructed the commission "as a first priority, [to] direct its attention to consideration of: protection of the rights and welfare of human research subjects; and issues in the management and use of genetic information, including but not limited to, human gene patenting."[9] In addition, however, President Clinton made two specific requests for reports and recommendations from NBAC following the announcement of significant scientific and technological breakthroughs. First, after the announcement that a sheep, Dolly, had been cloned, President Clinton requested that NBAC address within ninety days the ethical and legal issues that surround the subject of cloning human beings.[10] Then, following the announcement of the identification and isolation of human embryonic stem cells, President Clinton also requested a report and recommendations regarding the use of federal funds for such research. In addition, NBAC had the option of addressing other unspecified issues in light of several criteria:

1. the public health or public policy urgency of the bioethical issue;
2. the relation of the bioethical issue to the goals for federal investment in science and technology;
3. the absence of another or public body able to deliberate appropriately on the bioethical issue; and
4. the extent of interest in the issue within the federal government.[11]

NBAC's reports met, to varying degrees, all of these criteria.

How did the *Belmont Report* Influence NBAC's Thinking?

At the March 2, 1999, meeting of NBAC, Commissioner David R. Cox made the following comment regarding the commission's ongoing work: "And as for myself, . . . for any grounding on this I go back to the *Belmont Report*, and I [ask] what are the three components that we are talking about in terms of ethical responsibility in conducting research?"[12] Dr. Cox's statement reflects an important perspective in bioethics — and one of the lasting impacts of the *Belmont Report* — that the use of ethical principles (whatever their origin) may have relevance for individuals and/or public policy in working through particularly nettlesome problems.

Many thoughtful individuals find that referring to the *Belmont Report*, or indeed to any of the successive editions of Beauchamp and Childress's landmark *Principles of Biomedical Ethics*,[13] provides them with certain skills or helpful ways of thinking. It is another matter, however, as to whether certain groups of individuals — for instance, Institutional Review Boards (IRBs) or bioethics commissions — make similar use of these principles.

Among the tasks assigned to NBAC by the president's executive order was that the commission "shall identify broad principles to govern the ethical conduct of research, citing specific projects only as illustrations for such principles."[14] No direct reference is made to the principles described in the *Belmont Report* — respect for persons, beneficence, and justice — or, for that matter, to any other set of ethical, legal, political, economic, or scientific principles. This charge is not unlike the one given to the National Commission. As noted in the *Belmont Report*:

> One of the charges given to the National Commission was to identify the basic ethical principles that should underlie the conduct of biomedical and behavioral research involving human subjects and develop guidelines which should be followed to assure that such research is conducted in accordance with those principles.[15]

As we will show in the following section, the influence and relevance of the *Belmont Report* to NBAC's work and the particular use of ethical principles for addressing bioethical problems differed depending on the problem.

Research on Persons with Mental Disorders That May Affect Decisionmaking Capacity: An Example of the *Belmont Report's* Influence

NBAC recognized that this report stood "in a long line of statements, reports, and recommendations" and that much had changed since the National Commission completed its work. In our view, NBAC committed itself to preparing a report that followed in the tradition of the National Commission, both politically and methodologically: politically, because NBAC realized that the failure of the federal government to enact specific regulations for the protection of individuals with mental disorders left a gap in the oversight system of federal protections that needed filling,[16] and methodologically, because NBAC found much in the work of the National Commission that was relevant. Early in its report, NBAC writes that

> [our] views about respect for persons, beneficence, and justice are squarely in the tradition established by the National Commission nearly

20 years ago. Yet the research environment has changed, including the way in which research is conducted, its funding sources, and, in many instances, its complexity.[17]

And later, when discussing the role of advance planning, surrogate decision making, and assent or objection to research, NBAC notes that it

> agrees with the National Commission's conclusion in the Belmont Report that respect for persons unable to make a fully autonomous choice "requires giving them the opportunity to choose to the extent they are able" whether or not to participate in research.[18]

The report then makes clear its commitment to the *Belmont* principles, testing their relevance and finding them to be helpful in explicating and justifying a number of the report's recommendations. Consider the following examples.

> Recommendation 3: An IRB should not approve research protocols targeting persons with mental disorders as subjects when such research can be done with other subjects.

This recommendation was constructed with the *Belmont Report*'s principles of justice and respect for persons in mind. Indeed, the accompanying justification for this recommendation follows:

> because of this population's potential vulnerability, we should prohibit research that can be conducted equally well with others. . . . First it is important, on the grounds of justice and fairness, to discourage any tendency to engage these persons in research simply because they are in some sense more available than others. Second, this prohibition would further reinforce the importance of informed consent in human subjects research. The principles of justice and respect for persons jointly imply that IRBs should not approve research protocols."[19]

In general, this report gives considerable weight to the principle of respect for persons and its several implications. For example, in the text following Recommendation 6 (informed consent), NBAC explains that "nothing in this report is intended to supplant existing regulation regarding consent waivers. . . . A third party, such as a relative or friend, may not override the informed decisions of capable people. This is an implication of respect for persons."[20] Similarly, in the accompanying text for Recommendation 7 (objection to participation), NBAC argues that "even when decisionmaking capacity appears to be severely impaired, respect for persons must prevail over any asserted duty to serve the public good as a research subject."[21] In the text following Recommendation 13 (surrogate decision making), NBAC argues that "the principle of respect for persons, including respect for their autonomous choices, extends to their choices made while capable to cover future periods of impairment or incapacity."[22] Recommendation 17 (involving subjects' family and friends) emphasizes that "communication should, of course, be limited by the ethical and legal requirements of respect for personal autonomy and medical confidentiality."[23]

In addition to applying the *Belmont* principles to the issues arising in this report, NBAC also paid particular attention to two National Commission reports, both as points of reference and as reminders of proposals that had yet to be implemented. In its *Report on Research Involving Those Institutionalized as Mentally Infirm*,[24] the National Commission defined its scope on the basis of a contemporary problem that had received considerable attention by Congress. NBAC's report retained this approach but expanded its scope to include all persons with mental disorders that may affect decision-making capacity, rather than limiting it to individuals who may be vulnerable due to institutionalization.

There was considerable public commentary about this report. Sixty of the 118 individuals who provided written comments on a public comment draft expressed some reservations about the scope of the report, the majority arguing that the scope was too narrow and should be broadened to include all persons who are decisionally or cognitively impaired. Interestingly, only two respondents suggested that the report should limit the scope of its discussion to that of the 1974 National Commission's "institutionalized" subjects.

NBAC considered making specific recommendations regarding assent and dissent that followed those of the National Commission, ultimately adopting a recommendation that gives priority to subject objection to participate in research, while acknowledging that it is not necessary to adopt all objections to participate as permanent.[25]

NBAC referred extensively to subpart D of the federal regulations for the protection of human subjects,[26] which derived directly from the National Commission's report on children.[27] These regulations provided a specific referent for NBAC's recommendations for regulatory reform in three ways. First, NBAC reviewed the National Commission's proposed regulatory provision for a special panel convened by the secretary of Health and Human Services (HHS) and used it as a template for our recommendation that the secretary establish a special standing panel to review protocols and develop guidelines for research involving greater than minimal risk that does offer the prospect of direct medical benefit. Second, NBAC reviewed the categories of risk found in these regulations and determined that the justification provided for adopting three categories of risk (minimal risk, minor increment over minimal risk, and greater than minimal risk) was not convincing — even though many researchers encouraged NBAC to adopt this three-part approach. Third, NBAC distinguished between research involving greater than minimal risk that offers the prospect of direct medical benefit to the subjects and research involving greater than minimal risk that does not offer the prospect of direct medical benefit to the subjects. In making this distinction, NBAC was not adopting verbatim the language and justification adopted in the federal regulations. Indeed, as Robert Levine has argued (including in his public comment to NBAC on this report), the additional protections for children enacted in regulation did not correspond directly to the National Commission's recommendations.[28]

While it would be stretching matters considerably to suggest that NBAC adopted the ethical framework made explicit in the *Belmont Report*, the product of NBAC's deliberations reflected the spirit of the National Commission's thinking. At the same time, NBAC's report fails to fill in what we believe are important gaps in the *Belmont Report* and the regulations that followed from it. In particular, our report does not reach those human subjects who are not covered under existing federal regulations; we failed to insist that all human subjects be provided common protections, irrespective of the source of funding. However, it is worth noting that one of NBAC's first public statements was a unanimous resolution in 1997 that "no person shall be enrolled in research without the twin protections of informed consent and independent review of research"[29]

In retrospect, we know that this resolution was not strictly applied in this area of research. A second gap we failed to bridge is between innovative therapy and research. The simple truth is that new clinical procedures are developed by trial and error both in clinical settings and in formalized research trials. To be sure, the clinical setting does not have as sharp a set of conflicts of interest, but we would suggest that too little attention is paid to the protection of experimental human subjects in the clinical venue. Indeed, as will be clear from our discussion of cloning human beings, this particular issue was at the center of one of NBAC's recommendations.

Cloning Human Beings: An Example of Little Influence

The report on the use of somatic cell nuclear transfer (SCNT) techniques (i.e., Dolly-type cloning) to clone human beings was developed under rather unusual conditions.[30] Time constraints simply did not allow for the level of reflection that many would have wished because of the public hysteria that followed Ian Wilmut's historic announcement. Perhaps such an atmosphere was inevitable when what most people thought of as an extravagantly imagined future suddenly seemed to be a looming reality. Moreover, quite aside from the particular issues that surrounded the potential use of SCNT cloning for human procreation, the success of the Dolly experiment also brought to the surface — and to the mix of emotions already boiling — a great deal of pent-up anxiety about the extraordinary pace of scientific discovery and technological development and their potential impact on long-standing practices, values, and cultural commitments. At first blush, the news excited some, terrified others, and generated a great deal of uncertainty in general. From the point of view of public policy, these developments were not innocuous: within a short period of time, a number of proposals for ill-thought-out legislation and/or regulations were made, and the president banned the use of federal funds for the purpose of cloning human beings using this new technique. Bioethical issues surrounding the creation of human beings have historically been deeply divisive (if not hopelessly polarized) in our country, and while ethical disagreements are hardly unusual in a society as pluralistic as ours, we had particular difficulty in sustaining thoughtful and respectful discourse in this area.

As NBAC began its deliberations, several important points became apparent. The legal issues were many and complex, but in the U.S. context the most important of these was the constitutional issue of whether the cloning of human beings in this new fashion would be considered a private matter (i.e., part of a right to reproductive freedom) by the courts and therefore beyond the reach of government laws and/or regulation. Despite compelling arguments in exceptional cases, using these new techniques to clone human beings did not seem to NBAC a useful solution to the problems of infertility. Thus, there was no urgent or widespread social need sufficient to justify rushing ahead at the time. Many of the ethical objections to cloning human beings in this new manner are speculative, because they require the imagining of a future quite beyond our experience. Most importantly, for the moment, attempts to create human infants by this new cloning technique presented sufficient uncertainties and dangers to the mother and to the developing fetus to make such a project scientifically *and* clinically premature. For the immediate future, therefore, it was at least arguable that existing rules and regulations — including professional guidelines — would, for safety reasons, not allow attempting to use this new technique for the purpose of creating a human infant. At the same time, however, given the grey and troubled area between innovative therapy and existing regulations that focus on research trials, it was not clear what all this scientific uncertainty might mean in the area of innovative therapy. Clearly this raised the *Belmont* issue of human subject protection.

Very few issues raised by the prospect of cloning human beings by this new technique are not raised also by some combination of assisted reproduction techniques and/or genetic engineering. As was the case in the Paul Ramsey/Joseph Fletcher debates in the early 1970s, thoughtful ethicists held a wide variety of frequently conflicting views regarding the ethical issues raised by these new scientific possibilities. Dan Brock's commissioned paper for NBAC made this point clearly, laying out what he took to be the moral arguments in support of and against human cloning.[31]

Thus, while the various philosophical approaches provided substantial inspiration and guidance to our discussions, we knew that we would not be able to arrive at a set of recommendations solely through a process of philosophical reasoning and deliberation. We would have to reach actual decisions in some other way — using supplementary resources — and we would have to accept the fact that we would not be able to fully resolve some of the more contested, nuanced, and difficult issues in so short a time. As a result, some tactical decisions would have to be made to help us conceptualize our problem in a manner that would enable us to identify aspects of the issue where resolution was achievable. In essence, this involved finessing and/or putting aside certain dimensions of the issue so that we could make progress on at least some important dimensions within the time allowed. Following is a brief outline of this triaging process.

First, NBAC did not give in-depth attention to the constitutional issues that would be raised by any attempts to enact public policies or regulations in this area. We left these important matters to the future deliberations of our judicial system or to some other mechanism.

Second, we set aside consideration of how our health care system might address any ethical or philosophical questions (e.g., issues of distributive justice) raised by the implementation of this new technique. Indeed, all considerations that might be seen as evolving from the principle of justice were put aside.

Third, we did not have time to examine the substantial research base dealing with behavioral variation in twin, family, and adoption studies that could have helped us understand better the meaning of self-identity and individuality. We put aside both the growing literature on the impact, if any, of various new forms of parentage (e.g., anonymous sperm donors) on identity formation, in addition to the many mostly speculative efforts made over time to understand the discordance between one's self-image and others' images of one, and the resulting impact of this difference on identity formation.

Perhaps the most sensitive and controversial issues we finessed were those surrounding the issue of human embryo research — a particularly contentious issue in the United States — and the treatment of nonhuman animals. Ironically, as we discuss in the following section, NBAC would be specifically asked by the president to revisit the issue of embryo research eighteen months after the cloning report was completed, in our report on stem cell research.

In the final part of this chapter we discuss some of the lessons taken from NBAC's experience in preparing reports on very challenging bioethical problems. But we wish to identify one immediate lesson from the experience of the cloning report: even with thoughtful people of good will and the best intentions, large and complex ethical questions that go to the heart of what it means to be human and how to deal with the constant challenge of avoiding the full subjugation of individuals to purposes of others do not lend themselves to resolution in ninety days. It is difficult enough to understand who we humans really are, let alone who we might, or should, become. Recognizing this, we adopted a two-pronged strategy of (1) focusing our deliberations and our report on those ethical questions on which we could approach meaningful consensus and (2) identifying broader ethical, legal, and policy issues for more widespread and more sustained discussion over a more extended period of time.

NBAC's most acute challenge was trying to balance efforts to explore and understand a diverse set of complex but intriguing issues with our commitment to the president to aid in the process of public policy formation on a rapid timetable. We were constantly aware of a tension between our quite natural search for an ethically comprehensive approach and what became our even stronger commitment to develop a set of recommendations that might aid in the formulation of useful and effective short-term public policies that were based on an honest analysis of the ethical arguments and that satisfied an important, though perhaps not complete, set of ethical principles. In other words, we decided to try to define a middle ground between ethical discussion unrelated to the formulation of public policy and a formulation of public policy unrelated to ethical considerations.

Most (perhaps all) commissioners were sensitive to both the difficulty and the potential inappropriateness of proposing legislation — particularly legislation that had an impact on "private" decisions — on the basis of a contested and unsettled set of ethical issues. Here again, however, we limited the basis of our legislative recommendations to somewhat less contested ethical areas where the imperatives of safety, appropriate clinical practice, and the time needed for further ethical reflections outweighed the benefits of proceeding with premature and risky clinical procedures. It is important to emphasize that commissioners believed that even if these safety concerns turned out to be no more serious than other currently practiced interventions or common fetal exposures caused, for example, by maternal abuse of drugs or alcohol, more time was necessary to understand the ethical issues. Nevertheless, there remained considerable disagreement among the commissioners regarding a call for the enactment of legislation.

Stem Cell Research: An Example of Mixed Influence

On November 6, 1998, James Thomson reported in *Science* that he and his colleagues had isolated human embryonic stem cells from "excess" embryos from fertility clinics, donated by patients.[32] A few days later, on November 10, John Gearhart and his colleagues reported in the *Proceedings of the National Academy of Sciences* that they had isolated human embryonic stem cells from the developing gonads of aborted fetuses.[33] On November 12, the *New York Times* reported that Advanced Cell Technology, of Worcester, Massachusetts, claimed to have made human cells revert to their primordial, embryonic state, from which all other cells develop, and then by fusing them with cow eggs, created a hybrid cell that may have largely human cell characteristics.[34] On November 12, the Biotechnology Industry Organization issued a press statement calling on President Clinton to seek advice from NBAC, and on November 14, President Clinton wrote to NBAC requesting that "the National Bioethics Advisory Commission consider the implications of such [animal-human fusion] research at your meeting next week and report back to me as soon as possible." In addition, the president requested that NBAC "undertake a thorough review of the issues associated with human stem-cell research, balancing all ethical and medical considerations."[35]

Unlike the report on *Research Involving Persons with Mental Disorders That*

May Affect Decisionmaking Capacity, which the commission took eighteen months to complete, and unlike the report on *Cloning Human Beings*, which was completed in ninety days, the stem cell project was intended to be finished six months after it was requested (and it was completed within ten months).[36] While the time frame did not directly influence the type of report we wrote, it did have an effect on how much the commission would rely on the values and methods of agreement they had already used and how much they would have to constructively craft together.

Some of the ethical concerns we identified early on in our stem cell project can be classified under the principles of the *Belmont Report*. President Clinton's letter stressed that there was reason to revisit issues that had previously been discussed by the NIH Human Embryo Research Panel: "While the ethical issues have not diminished, it now appears that this research may have real potential for treating such devastating illnesses as cancer, heart disease, diabetes, and Parkinson's disease." As later evidence suggested, the principle of beneficence would play an important role in the report and its recommendations. Interestingly, in a letter sent to the president and members of Congress, signed by thirty-three former Nobel Prize winners, not only are the benefits from human stem cell research described, but so too is the suggestion of possible harms from failing to conduct this research.

NBAC heard testimony from Dr. Harold Varmus, the NIH director, supporting a similar view.[37] This theme — of the potential benefits forgone from the failure to conduct research and of the possible harms that may result — could be broadly interpreted as following from the principle of beneficence, which in the *Belmont Report* includes nonmaleficence. A careful examination of this issue would likely take note of how arguments from potential benefit or harm contrast sharply with the principle of respect for persons, even though the National Commission did not apply these principles in this way. And yet, one need only recall Hans Jonas's argument to appreciate how relevant this contrast was for NBAC as it attempted to develop a balanced policy response on the stem cell issue. Jonas wrote,

> Progress is an optional goal, not an unconditional commitment, and . . .
> its tempo in particular, compulsive as it may become, has nothing sacred
> about it. Let us also remember that a slower progress in the conquest of
> disease would not threaten society, grievous as it is to those who have to
> deplore that their particular disease be not yet conquered, but that society
> would indeed be threatened by the erosion of those moral values
> whose loss, possibly caused by too ruthless a pursuit of scientific progress,
> would make its dazzling triumphs not worth having.[38]

NBAC was particularly interested in how the principle of respect for persons might inform the discussion, although it refrained from adopting a deductive approach. NBAC considered two issues central to the discussions of many previous panels, commissions, and advisory bodies addressing the subject of fetal tissue and embryo research: the moral status of the human embryo and informed consent. Like other commissions, NBAC commissioned scholarly papers from recognized experts in the field, some of whom strongly urged NBAC to address directly and make recommendations about the moral status of the human embryo.[39] The president's request permitted NBAC to publicly deliberate about a topic that has been as divisive and controversial as any this country has had on its domestic policy agenda.

Other groups have attempted, with little success, to reconcile their public policy responsibilities with the need to take a position on the moral status of the human embryo.[40] Perhaps this is a subject that does not easily lend itself to the deliberative, multidisciplinary, and public process required of bioethics commissions.

Because the ethical acceptability of conducting research involving human stem cells will inevitably raise questions about how informed consent will occur, consider the following questions that could be addressed in a consent form for use by researchers seeking to obtain permission to use either fetal tissue or "spare" embryos for the derivation of human stem cells:

1. Are the purposes of stem cell research (in general) fully described?

2. If a specific research protocol is being contemplated with stem cells obtained, is the protocol fully described?

3. What are the possible risks to the woman (or partner) from the procedure to obtain stem cells, and how will these be minimized?

4. Will the consent form clearly disclose that stem cell research is not intended to benefit them directly and that the only "potential benefit" may be to advance knowledge?

5. Is there a description of the way in which one can change one's mind from initially deciding not to permit research on fetal tissue and then later deciding to permit this research to occur, *but* that once consent to permit research is given, research may begin?

6. Is it clear that decisions to consent to the procedures to obtain stem cells will neither enhance nor change the quality of care they will receive (e.g., by receiving additional care or treatment or by altering the clinical procedures to be used), nor will choosing not to permit the use of fetal material in research adversely affect the quality of care they will receive (e.g., by receiving less care or treatment)?

7. Will individuals be informed that no medical or genetic information about the gametes, fetuses, embryos, or stem cells derived from these sources will be provided?

8. What known commercial benefits, if any, are expected to arise for the investigators seeking to obtain human stem cells?

9. Is the source of funding for research (public, private, public/private, philanthropic) disclosed?

10. What measures will be taken to protect the privacy and confidentiality of individuals who provide gametes, fetuses, or embryos?

The ethical justification for this set of questions would be very familiar to anyone constructing consent forms for use in early stages of (basic) research, where there is no prospect of direct benefit to the individuals whose consent is being sought: Are the disclosures sufficiently detailed to afford the individual an opportunity to choose?

Adapting the principle of justice may be somewhat more difficult. The *Belmont Report* and the guidelines for IRBs found in federal regulation apply the principle of distributive justice rather narrowly: the distribution of benefits and burdens in research and the fair selection of research subjects. This limitation has been pointed out elsewhere,[41] but we wish to draw attention to one of the ways the principle might be used. One of the principal ethical considerations currently occupying health policy discussions is how best to allocate health resources — how, in other words, to fairly distribute limited or scarce resources, to provide benefits when not all can benefit. A number of allocation issues arise in research involving human stem cells: Should fetal tissue or human embryos — as potential sources for obtaining human stem cells — be considered a scarce or limited research resource? If so, what policy approach should be adopted for determining how to allocate or distribute this source? Can distributive justice explain how best to allocate such resources? Should those resources about which there may be greater or lesser moral concern be used irrespective of these concerns? For example, if it is argued that using tissue obtained from aborted fetuses is less problematic than using stem cells derived from embryos created specifically for research purposes, does this create a de facto priority of utilization? Are all resources of equivalent utility? Should, for example, resources that exist in abundance be shared more widely and used for more types of research than those that are not in abundance?

Some of the ethical concerns NBAC identified may not be captured easily under the principles of the *Belmont Report*. For some people, central to the discussion about the appropriateness of using federal funds to conduct research on human fetal tissue is the possibility of complicity in the procurement of stem cells. Is there a morally relevant difference between the derivation of stem cells (i.e., obtaining stem cells from cadaveric fetal tissue or from embryos in excess of those needed for in vitro fertilization [IVF] treatment) and using those same cells (knowing they had been obtained from these sources)? Is a researcher complicit in the abortion that resulted in the fetal tissue now being used for research? Clearly, the *Belmont Report* did not anticipate such questions.

The *Belmont* Principles in NBAC's Other Reports on Research Involving Human Participants

Not surprisingly, NBAC appealed to the *Belmont* principles both more broadly and more specifically in its three other reports on research involving human participants. In its report on *Research Involving Human Biological*

Materials: Ethical Issues and Policy Guidance, NBAC devoted one chapter to "Ethical Perspectives on the Research Use of Human Biological Materials."[42] This chapter uses the three *Belmont* principles as the primary but not exclusive framework of analysis. In addition to these principles, the report draws on "ethical guidance" in the federal regulations, rules pertaining to privacy and confidentiality, and perspectives offered by bioethicists and others, especially but not only in moving beyond individualistic interpretations of the principles and rules in order to "address the needs of relevant groups and communities." The report continues:

> Although NBAC in this report considers all of these sources of ethical guidance, it does not assume that they are equally authoritative or insightful. Rather, NBAC provides its own analysis of the major ethical issues and argues for specific ways in which to address the relevant moral concerns. Part of this analysis considers the extent to which research using human biological materials falls under the ethical principles and rules that ordinarily govern research with human subjects and the extent to which it is distinctive.[43]

In short, along with other sources of "ethical guidance," NBAC used and interpreted, but did not slavishly apply, the *Belmont* principles.

The *Belmont* principles also figure significantly in NBAC's report on *Ethical and Policy Issues in International Research: Clinical Trials in Developing Countries.*[44] This report grew out of the controversies surrounding research conducted or sponsored by the U.S. government or U.S. companies in developing countries, such as the placebo-controlled trial of reduced doses of AZT to prevent maternal-to-infant HIV transmission. While noting that "the principles embodied in the Belmont Report . . . serve as the foundation for the substantive ethical requirements incorporated into the system of protection of human research participants in the United States,"[45] the NBAC report recognized the complexity of their application to research in developing countries.

Justice was discussed in three important and rather unique ways. NBAC asked what the justification is for carrying out research in a developing (host) country when that same research could be conducted in a developed country. NBAC's answer appears in Recommendation 1.3 of the report: "Clinical trials conducted in developing countries should be limited to those that are responsive to the health needs of the host country." NBAC recognized, as did many other groups, that the conduct of research in countries whose collective health status is poor requires a justification that extends beyond a simple risk-benefit assessment. As one expert testified before NBAC, "Research should only be conducted in a country if the results will potentially directly benefit the population."[46] In this way, NBAC gave voice to need as an important notion of distributive justice. Need is, of course, among the more important notions of distributive justice,[47] but it has seen limited application in discussions of research involving human subjects.

The second way in which NBAC focused on justice was in asking what the obligations are, if any, to persons and communities after a trial is completed. NBAC gave two answers to this question, the first of which appears in Recommendation 2.2:

> Researchers and sponsors should design clinical trials that provide members of any control group with an established effective treatment, whether or not such treatment is available in the host country. Any study that would not provide the control group with an established effective treatment should include a justification for using an alternative design. Ethics review committees must assess the justification provided, including the risks to participants, and the overall ethical acceptability of the research design.

The argument for providing particular posttrial benefits was one of the most critically reviewed of any in this report, in part because of the difficulty in stipulating who would be entitled to potential benefits and on whose shoulders the obligations to provide those benefits rested. NBAC recognized that this issue implicated several *Belmont* principles, including justice. For example, we recognized

> the "responsive-to-needs" requirement as a manifestation of the core ethical principles of beneficence and respect for persons. The justification for requiring that research be responsive to the health needs of the population involved in it also rests on a concept of justice. . . . In conjunction with its discussion of justice and the distribution of the benefits and burdens of research, the Belmont Report touches indirectly on the issue of making effective interventions available to those populations upon which they were tested.[48]

NBAC then quotes the relevant portion of the *Belmont Report*:

> Whenever research supported by public funds leads to the development of therapeutic devices and procedures, justice demands both that these not provide advantages only to those who can afford them and that such research should not unduly involve persons from groups unlikely to be among the beneficiaries of subsequent applications of the research.

NBAC thus interprets the *Belmont Report*'s conception of justice as encompassing "the prospect of making effective interventions available to a population that is larger than that of the research participants, whether the population is a poor group within a wealthy society or one that lives in a developing country."[49] The second way in which NBAC spoke to this issue is in Recommendation 4.1:

> Researchers and sponsors in clinical trials should make reasonable, good faith efforts before the initiation of a trial to secure, at its conclusion, continued access for all participants to needed experimental interven-

tions that have been proven effective for the participants. Although the details of the arrangements will depend on a number of factors (including but not limited to the results of a trial), research protocols should typically describe the duration, extent, and financing of such continued access. When no arrangements have been negotiated, the researcher should justify to the ethics review committee why this is the case.

Two arguments were given for providing treatment to participants after a trial: one related to the unique features of the researcher-subject relationship, and the other related to justice as reciprocity — in which justice is concerned with what people deserve as a function of what they have contributed to an enterprise or society. NBAC argued that "justice as reciprocity could mean that something is owed to research participants even after their participation in a trial is ended, because it is only through their acceptance of risk and inconvenience that researchers are able to generate research findings necessary to advance knowledge."[50] As for providing benefits to those community members who did not participate in the study directly, NBAC still acknowledged a responsibility, based on justice, for research proposals to "include an explanation of how new interventions that are proven to be effective from the research will be made available to some or all of the host country population beyond the research participants" (Recommendation 4.2).

Hard questions remain, however, including who should seek to ensure justice in the provision of posttrial benefits and what level and kind of benefits are required for justice. These questions were partially captured by NBAC's discussion about capacity-building mechanisms — the third way in which the commission made use of the principle of justice. NBAC identified several strategies that could be employed by more economically developed countries to help less economically developed countries prosper and enhance and sustain their own research infrastructure. Seven of NBAC's recommendations focused on this broad set of issues, indicative of the importance the commission placed on developing and enhancing partnerships between institutions and countries. These included increasing resources for administrative and operational costs, funding education and training programs, establishing common approaches to ethics review by both host and sponsoring country IRBs, and using the equivalent protection provision of U.S. regulations.[51]

Ethical and Policy Issues in Research Involving Human Participants

In its final report, *Ethical and Policy Issues in Research Involving Human Participants*, NBAC focused on standards and procedures for research in the United States.[52] In addition to reprinting the whole *Belmont Report* in an appendix to the report, NBAC stressed that its "underlying principles . . . have served for over 20 years as a leading source of guidance for the ethical standards that should govern research with human participants in the United States."[53] However, these principles, and the regulations based on them, have not achieved the goal of fully protecting research participants, and "additional concerns" have arisen during these decades. Furthermore, some principles need clarification in light of various concerns and problems in interpretation and implementation. For instance, NBAC attempted to clarify risk-benefit analysis, as an expression of beneficence, by developing a component-based framework in order to enable investigators, IRBs, and potential participants to make better judgments.

What Lessons Can Be Taken from NBAC's Experience?

Contemporary and historical accounts of the National Commission provide an important starting point for any reflective assessment of NBAC's work. Most familiar has been the account provided by Jonsen and Toulmin, both jointly[54] and individually,[55] which suggests that the National Commission used a casuistic method of moral reasoning in which agreement was reached by commissioners on particular cases or problems, rather than by deductively arriving at conclusions from an existing set of principles. Beauchamp and Childress have questioned this interpretation, relying in part on an evaluation of the transcripts from National Commission meetings. They write, for example, that "the transcripts from the commission's deliberations show a constant back-and-forth movement from principle to case, and from case to principle."[56] Having read only the accounts of the National Commission's work, we will not presume to interpret which of these accounts is correct, although like Beauchamp and Childress, we suspect that while "evidence supporting Jonsen and Toulmin's interpretation exists . . . equally weighty evidence supports a justificatory role for principles in the National Commission's deliberations."[57]

Like other commissions, NBAC was challenged by the requirements of conducting its work in public and of constructing positions and arguments in territory fraught with disagreement and, occasionally, confusion. For example, in the period following NBAC's *Report on Cloning*, commissioners sometimes explained their rationale for agreeing with the report's recommendations by giving certain arguments more or less weight. Elsewhere, evidence suggests that individuals rely on different criteria for making ethical judgments in IRB decisions.[58] Hence, it might not be surprising to find that commissioners would describe their experiences differently. At the same time, we wonder how those observing NBAC's deliberations would assess its method.

Different bioethical problems may benefit from different approaches. This is not an especially new finding: the Advisory Committee on Human Radiation Experiments learned the same lesson and adopted a multimethodological approach.[59] The reports discussed in this chapter reveal that NBAC did not adopt an approach as much as it focused on a set of questions that it believed required answering.

Reports were not delayed by the lack of a formal ethical framework. Descriptions of how the National Commission worked through its reports vary — some claim that it started from principles and deduced moral conclusions for certain cases, while others claim that it began with cases and induced moral principles as presented in the *Belmont Report*. NBAC did not deliberately adopt either approach. Trying to develop consensus around a particular ethical framework among commissioners is extremely difficult. For this reason, consensus was a goal, but not at all costs.

The public and scientific communities think very pragmatically: How will proposed recommendations affect them directly? This model of pragmatism influenced by John Dewey has been observed in other commissions,[60] as well as in other areas of bioethics.[61] Critics of NBAC reports, arguing either on the grounds that the reports appear to lack sufficient ethical justification or on the grounds that they lack the empirical justification for making recommendations to reform federal regulations, will be unsatisfied with the model of pragmatism for such commissions, particularly commissions established by the executive branch. However, a bioethics commission appointed by the president, regardless of its independence in structure, function, or funding, will be perceived as an instrument of public policy that is pragmatic at its core, as well as an instrument of the executive branch. The commissioners were well aware that there was a wide spectrum of possible motives for the establishment of the NBAC. We always behaved as if the motive was to solicit expert advice, but we knew that there were likely other less flattering motives (e.g., to appear to be acting without actually acting). Thus, in addition to our pragmatism, we took the long view that if we could launch some useful ideas, they would have their impact in the fullness of time.

Finally, it is worth noting that NBAC discussed the possibility of developing a *"Belmont Report"* of its own — updating, if necessary, the National Commission's work by drafting a set of ethical principles that would be relevant and useful for researchers. This topic was discussed early in NBAC's life and revisited during at least two commission meetings. After considerable discussion, the commission decided that it could not devote time to such a project, in light of the uncertainty about when NBAC would cease to exist and the other urgent issues it needed to address.

Notes

1. U.S. Congress, Office of Technology Assessment, "Biomedical Ethics in U.S. Public Policy" — Background Paper, OTA-BP-105 (Washington, DC: U.S. Government Printing Office, June 1993).

2. Executive Order no. 12975, §3(a), 60 Fed. Reg. 52, 063 (1995). For an early history of NBAC, see Alexander M. Capron, "An Egg Takes Flight: The Once and Future Life of the National Bioethics Advisory Commission," *Kennedy Institute of Ethics Journal* 7, no. 1 (March 1997): 63–80.

3. 5 U.S.C. app. (1998).

4. 41 C.F.R. 101-6.1002 (c) (1998). The requirements state that "an advisory committee shall be fairly balanced in its membership in terms of the points of view represented."

5. 41 C.F.R. 101-6.1002 (d) (1997). The requirements state that "an advisory committee shall be open to the public in its meeting except in those circumstances where a closed meeting shall be determined proper and consistent with the provisions in the Government in the Sunshine Act."

6. Agencies may promulgate regulations that limit public comments at advisory committee meetings. 41 C.F.R. 101-6.1021 (d) (1998).

7. 5 U.S.C. app. § 9 (b) (1998).

8. Eric M. Meslin, "Engaging the Public in Policy Development: The National Bioethics Advisory Commission Report on Research Involving Persons with Mental Disorders That May Affect Decisionmaking Capacity," *Accountability in Research* 7 (1999): 227–40.

9. Executive Order no. 12975.

10. Letter from Harold T. Shapiro to President Clinton, November 20, 1998.

11. Executive Order 12975.

12. NBAC transcript, March 2, 1999, 50.

13. Tom L. Beauchamp and James F. Childress, *Principles of Biomedical Ethics*, 5th ed. (New York: Oxford University Press, 2001).

14. Executive Order 12975.

15. *Belmont Report*, p. 2, summary.

16. See, e.g., James F. Childress, "The National Bioethics Advisory Commission: Bridging the Gaps in Human Subjects' Research Protection," *Journal of Health Care Law and Policy* 1, no. 1 (1998): 105–22.

17. National Bioethics Advisory Commission, *Research Involving Persons with Mental Disorders That May Affect Decisionmaking Capacity* (Rockville, MD: NBAC, 1998), 9. The executive summary of this report is also found in *Federal Register* 64, no. 33 (1999): 8376–80.

18. Ibid., 27.

19. Ibid., 55.

20. Ibid., 57.

21. Ibid., 58.

22. Ibid., 61.

23. Ibid., 63.

24. National Commission for the Protection of Human Subjects of Biomedical and Behavioral Research, *Report and Recommendations: Research Involving Those Institutionalized as Mentally Infirm*. DHEW Publication No. (OS) 78-0006, 1978.

25. NBAC, *Research Involving Persons with Mental Disorders That May Affect Decisionmaking* , 29.

26. 45 C.F.R. 46. Subpart D–Additional DHHS Protections for Children Involved as Subjects in Research.

27. National Commission for the Protection of Human Subjects of Biomedical and Behavioral Research, *Report and Recommendations: Research Involving Children.* DHEW Publication No. (OS) 77-0004, 1977.

28. Letter from Robert J. Levine to Harold T. Shapiro and Eric M. Meslin, August 3, 1998.

29. National Bioethics Advisory Commission, *Ethical and Policy Issues in Research Involving Human Subjects* (Bethesda, MD: NBAC, 2001), 145.

30. National Bioethics Advisory Commission, *Cloning Human Beings,* 2 vols. (Rockville, MD: NBAC, 1997).

31. Dan Brock, "An Assessment of the Ethical Issues Pro and Con," in *Cloning Human Beings,* vol. 2, *Commissioned Papers* (Rockville, MD: NBAC, 1997).

32. James A. Thomson et al., "Embryonic Stem Cell Lines Derived from Human Blastocysts," *Science* 282 (1998): 1145–47.

33. Michael J. Shamblott et al., "Derivation of Pluripotent Stem Cells from Cultured Primordial Germ Cells," *Proceedings of the National Academy of Sciences* 95 (1998): 13726–31.

34. Nicholas Wade, "Researchers Claim Embryonic Cell Mix of Human and Cow," *New York Times,* November 13, 1998, A1.

35. Letter from President Clinton to Harold T. Shapiro, November 14, 1998.

36. National Bioethics Advisory Commission, *Ethical Issues in Human Stem Cell Research* (Rockville, MD: NBAC, 1999).

37. NBAC transcript, January 19, 1998.

38. Hans Jonas, "Philosophical Reflections on Experimenting with Human Subjects," *Daedalus* 245 (1969), reprinted in Jay Katz, *Experimentation with Human Beings* (New York: Russell Sage, 1972), 148.

39. See, e.g., John Fletcher's commissioned paper for NBAC; the testimony of Erik Parens before NBAC, January 18, 1999.

40. See, e.g., Lori Knowles's commissioned paper for NBAC.

41. See, e.g., the chapter by Patricia King in this volume.

42. National Bioethics Advisory Commission, *Research Involving Human Biological Materials: Ethical Issues and Policy Guidance* (Rockville, MD: NBAC, 1999).

43. Ibid., 42.

44. National Bioethics Advisory Commission, *Ethical and Policy Issues in International Research: Clinical Trials in Developing Countries,* vol. 1, *Report and Recommendations* (Bethesda, MD: Author, 2001).

45. Ibid., Executive summary, ii.

46. Ibid., 8.

47. Michael Walzer, *Spheres of Justice* (New York: Basic Books, 1983); L. Doyal and I. Gough, *A Theory of Human Needs* (New York: Macmillan, 1991).

48. NBAC, *Ethical and Policy Issues in International Research,* 63.

49. Ibid.

50. Ibid., 59.

51. 45 C.F.R. 46.101(h).

52. NBAC, *Ethical and Policy Issues in Research Involving Human Participants.*

53. Ibid., ii.

54. Albert R. Jonsen and Stephen Toulmin, *The Abuse of Casuistry: A History of Moral Reasoning* (Berkeley: University of California Press, 1988).

55. Albert R. Jonsen, *The Birth of Bioethics* (New York: Oxford University Press, 1998); Stephen Toulmin, "The National Commission on Human Experimentation: Procedures and Outcomes," in *Scientific Controversies: Cases Studies in the Resolution and Closure of Disputes in Science and Technology,* ed. H. Tristram Engelhardt Jr. and Arthur L. Caplan, 599–613 (Cambridge, UK: Cambridge University Press, 1987).

56. Beauchamp and Childress, *Principles of Biomedical Ethics,* 98.

57. Ibid., 95.

58. Eric M. Meslin, "Judging the Ethical Merit of Clinical Trials: What Criteria Do Research Ethics Board Members Use?" *IRB: A Review of Human Subjects Research* (May–June 1994): 6–10.

59. Advisory Committee on Human Radiation Experiments, *The Human Radiation Experiments* (New York: Oxford University Press, 1996). See also Eric M. Meslin, "Adding to the Canon: The Final Report," review of *The Final Report: White House Advisory Committee on Human Radiation Experiments, Hastings Center Report* 26 (September–October 1996): 34–36.

60. Jonsen, *Birth of Bioethics,* xx.

61. Glenn McGee, ed., *Pragmatic Bioethics* (Nashville, TN: Vanderbilt University Press, 1999).

6 The Principles of the *Belmont Report*

How Have Respect for Persons, Beneficence, and Justice Been Applied in Clinical Medicine?

Eric J. Cassell

This discussion[1] is framed by two anecdotes separated by forty-four years. In the spring of 1954, a man in his fifties was admitted to the teaching hospital of a medical school because he had developed symptoms of a heart attack a few hours earlier. He was selected to be the subject of the first attempt to use an innovative treatment (intravenous streptokinase and streptodornase) to dissolve the presumed clot of his coronary artery thrombosis.

The patient was chosen specifically because he was a derelict with no living relatives. In the fashion of the day, he was not told what was to be done and no consent was requested or obtained. No facilities existed at that time for the continuous monitoring of his heart, so devices were jerry-rigged to permit multiple intermittent recording of his electrocardiograms. An attending physician, resident, and medical student were in constant attendance. After a number of hours of receiving the new medication, an irregularity of his heart rhythm developed. The treatment was stopped out of fear for his safety.

The second instance is that of a thirty-eight-year-old woman with stage 4 (metastatic) cancer of the breast who, in January 1997, after almost three years of continuous disease and multiple treatments, was accepted into the bone marrow transplant program of a major western medical center. She was given high-dose chemotherapy followed by a bone marrow stem cell transplant. Months later a routine CT scan was done in the bone marrow transplant center of a distant city. It revealed what appeared to the transplant oncologist to be a recurrence of cancer in the spine. The implication was that the chemotherapy and bone marrow transplant had failed.

In 1998, the transplant oncologist sent the following letter:

Dear Olga, Cheryl, and Jimmy:

Enclosed is the relevant bone window from Olga's 11-12-97 CT Scan (as well as the formal reading) demonstrating the new sclerotic focus in the left pedicle of L2. I have circled it in red. It looks real to me and I would have Cheryl buzz [radiate] that area.

Olga, this is our only copy so will you send that one sheet back to us for our files? Hope all is well with the three of you. Talk to you soon.

Sincerely

[Signed]

Associate Director, Bone Marrow Transplant Program

Olga is the patient, Cheryl is the radiation oncologist, and Jimmy is the chief of the breast service at a major eastern cancer center in the city in which she resides.

In the years between these events, the following changes occurred in medicine:

- Chronic diseases became overwhelmingly the most common cause of death and reason for medical care, displacing infections and other acute diseases.
- Access to health care came to be considered a right, and ideas of distributive justice were commonly applied. Most Western nations (but not the United States) provided universal access to care.
- The therapeutic revolution took place grounded on progressively greater knowledge in medical science. Technological advance became a driving force in medicine.
- The cost of medical care rose progressively worldwide. Economic and legal forces became increasingly important in all facets of medicine and health care, frequently displacing moral determinants.
- The organization and financing of the delivery of medical services changed. Fee-for-service medicine withered, and physicians increasingly became employees of medical care organizations, were paid according to predetermined fee schedules, or received a capitated rate. Their political and social power shrank.
- Physician performance was increasingly measured by evidence-based process or outcome guidelines.
- The bioethics movement started in the 1960s and became a progressively more influential voice throughout medicine.
- The relationship between patient and physician shifted. Consumerism and ideas such as patient-centered medicine became commonplace. The public became very knowledgeable about medicine and medical science.
- With a few exceptions, the form and content of medical education changed little except that advances in medical science continually informed the curricula.

All these events and changes (among others) in medicine and medical care occurred during the same period in which the surrounding society was also in flux.

- The social unrest and antiwar protests of the 1960s challenged the hierarchical and social structure of the nation and were accompanied by a decreased respect for government and authority in general.
- Rights movements came to prominence, including the civil rights, women's, patients', disability, and gay rights movements.
- The metaphor of the United States as a melting pot was replaced by the rise of pride in ethnicity and diversity and a further increase in individualism.
- The ubiquity of computers and, latterly, the Internet produced a democratization of information and knowledge that was previously available only to professionals.
- The power of the law (in many forms) and money and financial incentives to determine social behavior and professional relationships increased, overwhelming the moral order. The bottom line became the bottom line.
- The gap between the rich and the poor grew steadily.

In clinical medicine the principles of the *Belmont Report* both influence and are influenced by the actions of individual patients and individual doctors and their relationships, even when the principles may be expressed by specific institutional procedures. For example, informed-consent documents and the laws and court rulings concerning informed consent are evidence of the widespread recognition of the importance of freedom of choice. However, informed consent was neither sought nor considered for the derelict patient who suffered a heart attack in 1954. Similarly, the principles of beneficence and respect for persons are institutionalized in hospital functions that monitor quality of care, such as the tissue committees that ensure that surgery is done for appropriate indications. The relationship of these formal institutional requirements to the expression of the principles is obvious, but patterns of practice, professional ideals, and the everyday behavior of both doctors and patients also demonstrate the definitions and application of the principles. They show what patients expect or demand and what physicians feel obligated to do.

Beneficence

I have chosen to discuss beneficence first, because the place of respect for persons and justice in clinical practice is easier to see when beneficence is understood. Actions or behaviors that are beneficent or benevolent are those that are intended to do persons good, promote the well-being of individuals, or that actively protect others from harm.

In the early 1950s (as in previous millennia), benevolence was best (but not only) exemplified by lifting the burdens of disease, as when physicians made people better. In the 1950s, for example, disease manifestations were treated simply because they were there. Hernias, hemorrhoids that caused any trouble, and most varicose veins of the legs were surgically removed, as were many superficial tumors and abnormalities. Except for the infectious diseases where newly introduced antibiotics had their impact, cancer, stroke, heart disease, arthritis, and neurological and similar chronic disorders were resistant to treatment, but physicians did their best to make the sick person's life better. Because doctors often could not actively improve a disease, whether consciously or not, they often focused on the patients who had the disease.

By the late 1960s, this simple idea of being better was changing as ideas of disease changed. The important place of classical structural pathology was giving way to understandings of functional abnormalities and pathophysiology, the concatenation of bodily events that lead to and define the abnormal state as well as explain its manifestations. With these changing concepts came the beginnings of the idea that a person was better when a part of the person was better. In medical science, increasing primacy was given to research on mechanisms of disease, including molecular biology, emphasizing, if possible, the patient (person)-body duality. Newer diagnostic technologies that facilitated the study of the body and its parts in motion — for example, cardiac catheterization, treadmill exercise studies, pulmonary function tests, various scans, magnetic resonance imaging, and rapid blood analyzers — made it easier to think only of the parts of the patient.

With an increasingly active research establishment and a growing technology, medicine became more activistic. Pharmacological innovation produced legions of drugs that gave excellent symptom control for complaints such as migraine headaches, angina pectoris, asthma, and panic attacks. The old belief that one should treat the disease not symptoms gave way to the belief that if you could treat a symptom, you should. Surgeons and surgical subspecialties could improve things from joints to obesity in ways previously unknown. Plastic surgery responded to the idea of medicine as a fix-it profession with all kinds of repairs, from old scars to breasts to eyelids. Increasingly, the organ or the physiological function — gastrointestinal disorders, heart and lung diseases, metabolic disorders (diabetes, hypercholesterolemia), hypertension — became the locus of medical action. The focus became not the person, but the part.[2]

The old idea of benevolence as simply the cure of disease, the relief of its manifestations, or the support of the patient came under scrutiny, as did the benevolent view of physicians in general. For example, until the 1970s, hysterectomies were frequently performed for enlarged fibroid tumors of the uterus, which were blamed for many of the ills of women, including abdominal and back pain, weakness and fatigue, and abnormal vaginal bleeding. There was wide geographical variation in the prevalence of the operation, which suggested that reasons other than fibroids of the uterus were determinants of surgery. Hysterectomies and other surgical procedures became an issue for the women's movement of the late 1960s and 1970s. For example, some feminists denounced gynecologists and other physicians as agents of the cultural repression of women. The self-help book *Our Bodies, Our Selves* (1973) was published to promote what the authors saw as the need for women to take back their bodies from physicians, the "benevolence" of whose motives and actions came under increasing suspicion. Doing "what the doctor ordered" without questioning and out of respect for his or her benevolent authority had long been the mode when I went into practice in 1961, but was largely gone by the end of the decade and has not returned.

With increasing knowledge about science and medicine, the public accepted medical definitions of treatment, improvement, and cure as evidence of physicians' benevolence. It is, however, if one can imagine such an attribute, a disembodied benevolence. It is not doctors, one might guess from the attitude of the public, but their scientific knowledge and technologies that diagnose, treat, and cure diseases. Knowledge of medical science and information about medicine began to pervade the media. With the advent of the Internet, patients now have an ever-increasing array of options from which to choose, leading to a kind of evidence-based and guideline-driven "cafeteria medicine." Patients (sick or well), now at center stage in medicine, themselves define what will be read as benevolent, as physicians retreat from taking responsibility for the whole patient.

In the care of the dying, the paradoxes of beneficence are easily seen. The goal of keeping people alive first entered medicine in the nineteenth century, well before the necessary technical capability existed. By the third quarter of the twentieth century, however, resuscitation techniques and life-extending therapies made it possible to keep patients alive who would previously have died of their diseases. As time went on doctors became better able to support one physiological function after another, apart from the state of the whole patient. Kidney dialysis replaced lost renal function, better ventilators replaced failed lungs and supported oxygenation, pacemakers and defibrillators maintained heart rhythm, total intravenous nutrition took over when oral nutrition failed, various modes of blood pressure support and volume replacement maintained circulation, and transfusions of various blood components as well as means for stimulating production of blood elements allowed for the continued function of blood as an organ.

Even terminally ill patients unquestionably soon to die could be maintained by advanced life support systems. It came to be believed that remaining alive was the ultimate good, whatever the future might hold. On this basis, the judgment of physicians about which patients should be resuscitated was superseded by legal requirements that resuscitation be attempted for *all* patients whatever their cause of death.

By the 1980s, intensive care units contained patients on life support, though they had no chance of returning to meaningful life whatever the outcome of their therapies. These patients lay alongside others with diseases for which resuscitation and life support were appropriate, because, if they could be maintained long enough, return to full function was probable.

Medical beneficence was thus defined in technical terms, and "keeping someone alive at all costs" was redefined as a medical good. The effect on medical practice was significant; when codified definitions of benevolence were not met, patients were considered not only harmed but also wronged, in that their legal rights had been violated.

These excesses led to a reaction among the public and physicians. The importance of a good death, first brought to public awareness by Elisabeth Kubler-Ross in her book *On Death and Dying* (1972), received increasing support and was the subject of widespread discussion. Advance directives and "do not resuscitate" orders became more common, and the assignment of surrogates for medical purposes became easier and more frequent. The hospice movement provided an alternative for the care of the terminally ill and focused attention on the relief of pain. Nonetheless, at least in the hospital, technical proficiency continued to accept that good done to a part was as good done to a patient.

The environment of medical care had changed so that patients were consistently told what was wrong and what was happening in considerable technical detail, and then given technical options to choose from, as was the case with Olga, the patient with terminal breast cancer mentioned earlier. She, like others with similar end-stage diseases, chose to accept the physicians' recommendations, perhaps because few other choices were described to her.

The hospital care described in the previous paragraph is focused on acute events, so that chronic illnesses such as diabetes, coronary heart disease, stroke, cancer, AIDS, and neurological disorders are often treated as a series of acute episodes. From this perspective, it is possible to consider the abnormal process completely apart from the person in whom it occurs. The technical details of ventilatory support in lung failure are independent of the particular patient. Some patients tolerate the treatment better, complain less, and are less demanding and more cooperative than others, but these personal characteristics are not an important part of the clinical situation when it is defined in primarily technical terms. Acute events continue to be the major focus of clinical policy — the aggregate of medical education, hospital care, medical insurance, outpatient services, medical research, and all other facets of the institution of medicine — including, as we have seen, definitions of benevolence.

In the forty-four years between the cases that opened this chapter, the concept of benevolence in medicine has changed considerably. Initially, the idea of doing good and avoiding harm was seen as resulting from both physicians' personal characteristics and their ability to make patients better. The former, if ideal, would be responsible, trustworthy, careful, and devoid of overweening pride, venality, and impure motives. The latter was a function of technical knowledge and proficiency. The physicians of the derelict with the heart attack exemplified both moral and technical inadequacies of which they were largely unaware. They did, however, unquestionably know of the dangers to the patient and were fearful of harming him.

The intervening period in medicine has seen an explosion of technical capacity and a great increase in moral awareness. The concept of benevolence has shrunk pari passu. The personal characteristics of physicians that served benevolence and were believed to be of such importance in previous generations now serve nostalgia more than clinical medicine. The code of what was called medical *ethics* in times past was devoted to protecting patients (among other goals). Now termed medical *etiquette*, it has largely disappeared. The good of patients that was identified with making them better has changed as concepts of disease have evolved. A structural understanding of disease, best suited to an era of acute disease, has been superseded to a large extent by a pathophysiologic perspective that focuses on function. This paved the way to measuring benefit by good done to a part of the patient, or only one body system, while benevolence, measured by benefit to the whole patient, is more honored in words or advertising slogans than in medical actions. By the end of the period, contradictory trends emerged. The patient has increasingly become central, but codes, guidelines, laws, and legal actions have pushed the notion of wronging the patient to the fore, while the calculus of benefit and harm has receded as physicians have withdrawn even more from the ideals of the past.

Respect for Persons—Acknowledgment of Autonomy and Protection of Those in Whom It Is Determined

The physicians of the derelict with the heart attack probably did not entertain the notion that he had a *right* to decide whether to participate in the experiment or that he was *wronged* by not having been asked for his consent. They chose a derelict with no family because more sophisticated patients were always wary of being "experimented on." By the standards of the time they did the right thing: they protected the patient from harm. He was, after all, *a patient, not a person*. The letter to Olga, the young woman with stage 4 breast cancer, suggests that to the physicians Olga is clearly a person; it is the *sick patient* part of her identity that seems to have diminished. *She has gained rights as a person and lost obligations due a patient.*

The idea of respect for persons as described in the *Belmont Report* — or even the concept of persons qua persons — was not present in medicine at the beginning of the period covered by this chapter. Benevolence and the avoidance of harm were the expressions of respect for the humanity of patients. Patients were to be treated as fully human. Persons, on the other hand, are not merely human; historically they are philosophical, legal, and political entities with rights. Because of this, not only can persons be harmed, they can also be wronged.

Paul Ramsey's book *The Patient as Person: Explorations in Medical Ethics* was published in 1970. However, the directive "Treat the patient as a person" had been in use since the 1950s. Its literal meaning seems simple: treat a patient as if a patient were a person. In the early 1970s, this appeared so obvious as to be banal, but in fact, during that period the personhood of patients was still only emerging. Prior to that time they were *not* persons. When persons became patients, their social status changed. In the late 1960s, I admitted a mentally fit corporate president with pneumonia to the hospital. After I explained to him what I thought was wrong and what would take place, his wife and I went out into the corridor for a full discussion of his case — a discussion that would not now take place without his participation. Patienthood had in minutes deprived him of his status as a self-determined adult. This was the fashion of the times in the United States.

In fact, it seems probable that the idea of "person" as we use it today — derivative as it is from the evolving concept of atomistic individuality — was just beginning to take full form after World War II. In the time period covered in this chapter, the nature of persons changed, society changed, and medicine changed, and the result was a change in the meaning of respect for persons and autonomy.

In the 1950s and early 1960s, women in public were not persons in their present sense, nor were people with disabilities, nor gay people. The civil rights movement achieved legal rights for blacks and other ethnic minorities, but also changed their social status by making them persons in the wider American community in both legal and political senses. These changes were not the end of the matter; they were the official beginning of a process that had started well before the civil rights movement and that continues to this day.

In *The Patient as Person*, Ramsey also discusses the bond between physician and patient and how both are defined by that bond. Before the patient fully became a person, physicians were their decision makers. The doctor made decisions about the best thing to do and about what and how to tell patients about their circumstances. It was part of physicians' obligations and part of their patients' expectations. Good physicians knew that patients had to be informed about what was happening because too much uncertainty was considered bad. But full disclosure of fatal or dangerous diagnoses or situations was thought to be harmful because it would be followed by a sense of hopelessness. When a patient I took care of returned to his room after his surgery for inoperable cancer of the stomach, he asked the surgeon what he had found. The surgeon said, "We did a lot of cuttin' and schnitten and removed a lotta junk and you're gonna be fine." I took care of the patient until he died months later. A few days before he died, he said, "Sometimes lately I think maybe I'm not getting better."

Medicine was only a few decades into the beginning of the therapeutic revolution that is now taken for granted. Then, despite great expectations of the bounty from medical science, there was little optimism about the outcome of diseases such as cancer, strokes, heart attacks, heart failure, advanced diabetes, and emphysema. Only for children had everything improved, as their death rates dropped precipitously in the Western world because infectious diseases were now readily treatable.

It is important to understand the relation among the fall in death rates, the improvement in health and well-being, the optimism fed by scientific advances, and the notions of respect for persons and freedom of choice. Previously, if you believed that your cancer inevitably meant a hopeless outcome and a painful death, and if your physicians believed that there was nothing beyond surgery that could be done for you, you might not have been so eager for knowledge or the freedom to choose. Beyond refusing or agreeing to (say) surgery, there was not much choice. One did the radical mastectomy and waited for the patient to get a recurrence or be lucky. So doctors lied, not because they were morally defective, but because, in their eyes, all they had to offer was denial of a bad truth. Especially because, at that time, personal matters that might arise from these illnesses and the doctors' lies — for example, lost hopes, unhappiness, anxieties, sadness, suffering, death, and grief — were also private matters kept from the view of others, even physicians (unless they looked). If, on the other hand, death rates are falling, the expectation of becoming hopelessly ill is disappearing in the face of new treatments, persons with disabilities are enjoying active lives in increasing numbers, optimism pervades medicine, and the world around is encouraging a further blossoming of individualism. In this climate telling the truth and freedom of choice have new meaning.

The effects of the change in disease burden, the advance of medical science, the change in social status, and personal freedom are easily seen in the rise of the women's movement. Would the continuing emergence of women to their present social and political state have been possible without a low birthrate, effective contraception, the virtual disappearance of childbirth complications, and the increased survival of children? As recently as 1928, Virginia Woolf, in *A Room of One's Own*, could decry the paucity of women in letters or any other profession. At that time, none of the four previously mentioned benefits were available to women. Is widespread freedom of choice possible in their absence? Virginia Woolf did not think so. In the 1960s, however, these were the facts. World War II had put women in the workforce, as had World War I, but, I believe, it took these medical changes to continue their advance. By the end of the social turmoil of the 1960s, as the women's movement grew, abortion had become legal, common venereal diseases were easily treatable (although new ones were appearing), and the physical constraints on the emergence of women were disappearing. Further, the opinion held by physicians about women gradually changed as women changed their personal and political aspirations and their demands on medicine so that their climate of choice was also altered — even in advance of the entrance of large numbers of women into medicine. Earlier, I discussed the change in the attitudes of many women toward physicians and medicine as a result of ideas prominent in the women's movement. Women wanted to recapture (militancy intentional) their own medical care and make their own decisions particularly, but not exclusively, about contraception, sexuality, diseases of female genitalia, and pregnancy. They seized the locus of choice from physicians prior, I believe, to a similar change in the general population.

The bioethics movement was a major force in spreading the importance of patient autonomy in clinical medicine. Publications, public discussions, the education of interested physicians and individuals who were making bioethics their academic field, and increased public interest brought power to the idea of patient autonomy. Respect for persons in clinical medicine had become synonymous in many minds with autonomy defined solely as freedom of choice. For example, the protection of individuals with diminished autonomy seemed to be identified with the problem of the resuscitation of persons who had lost decision-making capacity. Issues such as the nature of person, the impact of illness on decision-making capacity, the problems of autonomy that were specific to medicine and care of the sick, and the meaning of autonomy in the context of the special relationship between patient and physician were buried under the tide of legal interpretations of these concepts and the rise of the language of rights. Two issues — autonomy and its apparent natural enemy, paternalism — were raised so often that, by the 1980s, Daniel Callahan could speak of the "desert of autonomy and paternalism" that had spread over bioethics. These two terms, in themselves essentially political, denied more complex or subtle interpretations of the actions of patient or physician or their relationship. For example, attempts by physicians to influence or take part in their patients' decisions were often tarred with the brush of paternalism.

As time went on, the emphasis in the meaning of freedom of choice in medical practice shifted from choosing among the reasonable alternatives offered by physicians to doing whatever the patient (or surrogate) wanted. This was most evident in intensive care units where unconscious patients with no possibility of survival in the absence of support equipment were kept alive because (the physicians said) the family wanted a "full court press." It was not unusual at this time for the family and the medical staff to become adversaries. Influential guidelines in the bioethics literature (e.g., Hastings Center Guidelines) supported the right of the family or patient to insist on resuscitation no matter what the clinical situation or the patient's prognosis. In earlier years, the art of prognosticating and basing clinical decisions on the probabilities of possible outcomes rather than solely on what the doctor wanted had been very important in clinical medicine. With freedom of choice, this element began to disappear from clinical medicine. (Surgeons remained constant in this regard. They remained firmly in control of the decision to operate — if the patient agreed.) Absent concern about the impact of the past and the future on a clinical decision (what prognostication is all about), the exercise of autonomy in medicine came to be marked by immediacy. As required by law, in every hospital in New York state, a "patients' bill of rights" was posted prominently next to elevators or other visible sites. The tone of the document was adversarial, as though everything that could be undertaken by physicians was determined by patient rights and medical obligations rather than patient needs and medical responsibilities. The balance of power had clearly shifted to the patient. In ambulatory, nonhospital settings, autonomy also increasingly meant simple freedom of choice in a medicine of the immediate. For many physicians it became easier to acquiesce when patients wanted (say) medications or diagnostic technologies than to assert medical authority or negotiate a middle ground. Patients and the public at large had become so knowledgeable that their choices often seemed well informed and cogent.

There can be no freedom of choice in the absence of knowledge on which to base the choice. As previously noted, until the 1970s, physicians commonly withheld the truth from patients who had life-threatening diseases. Earlier doctors did not tell patients about the facts of their illnesses even when they were not serious. For example, doctors frequently did not reveal blood pressure levels to patients. Why would they want to know? After all, doctors thought, they did not know what a specific blood pressure meant. All it might do is worry them if it seemed high. The reasons for a specific medication might be revealed because it had been shown that explanations increased compliance (otherwise only about half of prescriptions were ever filled), not because it was believed that patients wanted to participate in the decision. Why would they? That was the doctor's job. It was commonly believed that doctors did not tell the truth — that they hid bad news. By the late 1970s, patients were increasingly told the truth. By the late 1980s, any reticence on the part of the physician about revealing the truth was gone. The only criterion for telling something to a patient became its truth.

From the destructiveness of complete lies to the destructiveness of unmediated truth took less than three decades. Attempts to teach the harm that could be done by "truth bombs" and "truth fragments" fell on deaf ears. But the truth of information is only one of its aspects. Also of importance are accuracy, reliability, timing and completeness, the meaning to the patient of the information, its relevance to the patient's problem, whether the patient's illness has impaired the ability to understand, whether it increases or decreases uncertainty, what it indicates about appropriate or possible action, and what impact it has on the relationship between patient and physician. The understanding that information is a tool that can be used for healing or hurting disappeared under the new avalanche of truth revealed to patients in the service of autonomy. Currently medical students and house officers tell patients what is happening apparently without reservation whenever it arises and whatever the nature of the information. This is a task that, because of its delicacy, was previously reserved for the attending physician. When the delicacy disappeared, so did that remnant of etiquette. *When physicians tell all in the manner described, they separate themselves from responsibility in the relationship.* They are like a stock market ticker, the impersonal dispenser of whether the person's stock is rising or falling. Seemingly, they do not have opinions about what is good, bad, or indifferent; they just know the facts — and a treatment. Deciding what should be said when, where, and how requires knowledge of not just the medical facts but the nature of the sick person and his or her needs beyond the simply "medical." This kind of personal knowledge cannot be obtained by physicians who have distanced themselves from their patients.

Decisions made in the name of respect for persons and their autonomy can result in different conclusions about the right thing to do. Consider, for example, the following two cases: A terminally ill patient with respiratory disease decided against further treatment and entered a home hospice program. He soon became very sick and was brought to an emergency room in respiratory failure. Severely short of breath, he chose to go on a respirator despite having previously decided against resuscitation. His request was granted, although he could have been made comfortable without a respirator, and ultimately he will again be in the terminal state he was in before entering the emergency room. The second instance is that of a patient who had been on dialysis for a long time and decided to stop treatment. When he was close to death he requested that he be restarted on dialysis. His physicians chose not to do so, and he soon died.

In the first instance, the decision was justified by saying that the patient wanted to be resuscitated despite his previous refusal of further treatment. In the second instance, the decision was justified by saying that the patient's previous decision against dialysis carried more weight than his current request for dialysis. The first case probably represents the more common contemporary occurrence. Here the patient's choice is atemporal — as though the person of the past does not count in the present and as if there is no future. Choice is exercised as if it were independent of circumstances — as if the panic of respiratory distress had no impact on the choice and the patient in his profoundly sick circumstances is as representative of the person as the less sick voice of the recent past. It is the immediate choice that counts. In the second instance, the physicians take responsibility for deciding that the previous decision to stop dialysis is more representative of the person than the current choice to restart dialysis. How do they *know* that they are correct, that they are not condemning the patient to death based solely on their judgment? They cannot *know;* it is merely a judgment. On the other hand, their decision is based on their knowledge of end-stage renal disease, the life of a dialysand, and this patient's previous experience with both. The patient will die of renal disease — no action or decision will change that fact. His previous decision has duration, was made over time, and was justified over time. No patient is removed from dialysis without a lot of discussion with his or her physicians — it is in the nature of dialysis units. To honor his immediate choice would return him to the situation that he opted to end with death rather than face its continuance. Here the decision has duration, is not immediate, acknowledges the effect of illness, and, perhaps most important, entwines the acts of the physicians with those of the patient. The case of the man with renal disease treats choice in a more communitarian manner, with its stress on mutual respect. The case of the man with respiratory disease treats choice in a more individualistic manner — much as the law does.

In the period covered by this discussion, clinical medicine moved *away* from respect for the humanity of patients, expressed primarily by kindness, compassion, the avoidance of harm, and benevolence, with little regard for the rights of patients to make informed choices. It moved *toward* respect for persons defined by autonomous freedom of choice, with little regard for other aspects of autonomy. Before the current era, patients were not accorded full status as persons — sickness automatically removed them from the community of equals, impaired their autonomy, and required that physicians accept full responsibility for their benevolent treatment. At present, in the absence of obviously diminished mental capacity, the easily demonstrable impairment in the very sick of the ability to make reasoned decisions is turned aside and they are accorded the full autonomy of normal persons. This reduces the degree of responsibility of their physicians.

Justice

In 1981, I was asked to discuss justice as it applied at the patient's bedside. I argued that "love of humanity, compassion, and mercy, *not* justice, are the appropriate concepts to guide actions at the bedside."[3] In the years that have followed, this idea has been challenged. Problems of distributive justice — distribution of income, housing, or other societal goods — necessarily arise in a nation where a substantial minority of individuals lack health insurance. Justice, in these and other matters, is primarily a problem of large-scale social institutions such as governments. In recent decades, because of the advent of expensive technologies or treatments for which there are inadequate resources (such as organ transplants or gamma knife radiation units), medicine has been the arena of other problems of distribution that are the purview of more local institutions such as transplant teams, hospitals, or other medical care organizations. The rise of managed care in the last decades has highlighted the distributive issues that arise when cost becomes the primary value by which services are measured.

Inevitably this involves the belief that individual physicians should play a part in preserving society's medical resources. Simply put, this means that the physician should be thinking not solely about a particular patient but also about how the resources used in that patient's care affect conservation of the general resource supply. Only a few decades ago, such an idea would have met with strong opposition. The ethos held that physicians' primary obligation is to their patients, and all else comes second — including physicians themselves, their institutions, and society. The issue is not whether conflicts of interest exist, but whether there are durable values that protect the patient, such as the duty to the individual patient that resonates through medicine's history.

It is not surprising that in this changing climate attention has turned to issues of justice arising from the individual physician's attention to an individual patient. The idea of concepts of justice applying to the physician's acts at the bedside, to which I denied legitimacy in 1981, has now become a focus of attention. In 1992, Jon Elster published *Local Justice*,[4] which explores allocation of resources in situations not usually considered matters of justice. Examples including military draft, admission to colleges, and certain larger medical allocation problems such as organ transplantation are discussed at some length. Elster cites previous work by others, including Michael Walzer, that has focused on similar local issues. *Local Justice*, however, allows me to demonstrate the application of these ideas to clinical medicine. The values underlying Elster's arguments are simple. To meet the standard of justice, the distribution of scarce resources should be both equitable and efficient. The existing norms of clinical medicine appear to conflict with justice as a principle of clinical medicine. The following excerpt makes the point:

> In many cases, professional norms are self-explanatory. There is no need to ask why colleges want good students, why firms want to retain the most qualified workers, or why generals want their soldiers to be fit for combat. The norms of medical ethics, however, are somewhat more puzzling. I shall offer some conjectures concerning the origins of two central medical norms with important allocative consequences. Neither norm is outcome oriented, in the sense of aiming at the most efficient use of scarce medical resources. Instead, one might say that the norms are *patient-oriented*, in a sense that will become clear in a moment (italics in the original).
>
> The first is what I have called "the norm of compassion," that is, the principle of channeling medical resources toward the critically ill patients, even when they would do more good in others. In addition to spontaneous empathy, I believe some cognitive factors could be involved in this norm. . . . Instead of comparing the fates of different individuals if treated, doctors compare their fates if left untreated. . . .
>
> Next, there is what I shall call "the norm of thoroughness." Rational-choice theory tells us that when allocating scarce resources, whether as input for production or as goods for consumption, one should equalize the marginal productivity or the marginal utility of all units. . . . A rational consumer would, therefore, spread his income more thinly over a large number of goods, rather than concentrate it on just a few.
>
> We can apply similar reasoning to the behavior of doctors. With respect to any given patient, the doctor's time has decreasing marginal productivity, at least beyond a certain point. . . . This implies that if a doctor makes a very thorough examination of his patient, his behavior is not instrumentally rational with respect to the objective of saving lives or improving overall health. Other patients might benefit much more from the time he spends on the last and most esoteric tests. Nevertheless, doctors seem to follow a norm of thoroughness, which tells them

that once a patient has been admitted, he or she should get "the full treatment."

In Norway, a recent parliamentary commission found that eye specialists tend to admit too few patients and treat each of them excessively thoroughly. When I confronted my own eye doctor with this claim, she refuted it by telling me about a case in which she had been able to diagnose a rare eye disease only after exhaustive examination, thereby saving her patient's sight. I did not remind her of the cases that go undetected because the patient never gets to see a doctor at all.[5]

In the 1950s, such an application of economic theory to medicine was unlikely. Even today, many clinicians would be upset at the conclusions Elster has drawn, but he is not alone. As F. H. Bradley once said, "When you are perplexed, you have made an assumption and it is up to you to find out what it is." Elster's assumption, which led to his puzzlement about the norms of medical ethics and on which his argument stands, is that medicine is devoted to saving lives and promoting overall health. Historically, *clinical* medicine has been devoted to individual patients — one at a time. Elster can be excused his error. He had probably been reading medicine's public relations slogans, in common with the rest of the population, and the medical industry — clinical, teaching, and research — supports itself by spreading the belief that it is about saving lives and promoting health. The error is really an error in systems theory. The level of the medical system devoted to these goals is not medicine as a profession of individual doctors treating individual patients — what most people think of when they speak of medicine. It is medicine as a social system, concerned with keeping the population alive and healthy. The United States does not have an institution responsible for the social system of medicine — certainly not the Surgeon General's Office or the Department of Health and Human Services. The nation depends instead on the outmoded assumptions that the health of the population is the sum of the health of individuals and that lives are best saved by the actions of individual physicians — both demonstrably false.

In the last few decades, however, as the economics of medical care have come under increasing scrutiny — because of rapidly rising costs and their assumption by government and other economic entities — addressing questions of equity and efficiency in the care of patients has been seen as necessary and reasonable. However, the goal has not necessarily been the best medicine for the overall health of the population and the lowest death rate, but rather the most medical care for the money. Perhaps the closest thing to an arbiter of medicine as a social system has been the Healthcare Financing Administration in conjunction with various organizations concerned with technology assessment, epidemiology (The Centers for Disease Control), and health policy.

It is probably true that at the present time more than one set of norms are applied to clinical medicine and the care of patients (who persist in clinging to the belief that when they are actively the patient they are their doctor's primary concern). Their health insurance organization is probably dedicated primarily to efficiency and, hopefully, equity — whatever its public advertising may say. But the matter does not end here. A recent paper by Lynn Jansen, a nurse who has a doctorate in political theory, allows us to move a step further.[6] Based on Elster's work, she applies the concept of local justice to the treatment of pain. She finds, as have many others, that pain is undertreated. In her discussion, she states that "an important factor affecting the distribution of these [pain management] resources was the decisions made by individual clinicians at the bedside. Since these decisions affect the distribution of important health care resources, they should be understood as raising an issue of justice." After citing as an objection to her conclusions the belief of others that individual treatment decisions should be discussed in terms of beneficence, she states,

> It is the actual distribution of resources, however, that should be assessed in terms of justice. *Ultimately, what matters from the standpoint of justice is who actually gets what resources.* If, therefore, this distribution is influenced in part by the decisions of individual physicians, then it is entirely appropriate that these decisions be assessed in terms of justice (italics in the original).

(In a footnote she states that not every decision by a physician raises an issue of justice.) Why does it matter whether these local decisions are viewed in terms of justice?

> As the case of pain management resources aptly demonstrates, many of these resources cannot be distributed properly according to a uniform policy or guideline. Yet they are sufficiently important to require a stronger distributive justification than simply relying on market forces or professional discretion.

And finally, "When decisions . . . come to be viewed in terms of justice, there is greater pressure, both social and legal, for those who make these decisions to defend and justify them in public."

Whether one agrees with Jansen's argument is not the issue; what is important is the concept on which her discussion is based. For Jansen, and for many others in the last decades, the actions of the doctor have become *resources* for which physicians are socially and legally accountable. Take away the concept of resources, and the argument that the idea of justice applies at the bedside disappears. The overriding belief that physicians' acts represent the exemplification of the *personal* duties of individual physicians toward individual patients — that this is the moral framework of clinical medicine — has lost considerable contemporary currency. A number of things follow from the shift to a framework of justice. This shift presupposes people or groups pressing claims for scarce goods as their *right* and justifying them by rules or standards. It suggests the utility of rules and guidelines and evidence-based medicine that can provide the basis for the social and legal evaluation of the distribution of the physicians' resources. It provides a basis for diminishing the importance of the personal judgment of physicians.

In quoting both Jon Elster and Lynn Jansen, perhaps I have presented an overdrawn account of the place of justice in contemporary clinical medicine. This idea is not usually so baldly stated because, in a period of changing values, people often talk the old values but act on the new. Whether the word *justice* is used or not, the idea of physicians' services as scarce commodities discussed in marketplace terms is, by now, widely accepted. Similarly, the utility of guidelines, evidence-based medicine, and rules of practice are increasingly accepted. When physician commentators point out that such a medicine dismisses the importance of the physician's relationship with the patient and the importance of personal judgments based on the evaluation of the individual patient in context, they are correct, but they miss the point. To a medicine guided by marketplace principles and the socially based ethics of justice, the loss of the personal is irrelevant. The classical norms of clinical medicine — dedication to the patient, constancy, thoroughness, self-discipline, compassion — are not about saving lives (plural) and overall health, they are about *this patient's* life and health. These are values that occur within a framework that at the moment of action knows no other patient. It is the physician's difficult task, accepted since antiquity, to keep these values in the forefront despite the fact that at any time there are many other patients. *They are values that arise in relationships* — they presume a relationship between doctor and patient. In this relationship, fairness — justice — is only one duty among others and probably is not preeminent, judging from its absence in classical discussions of the obligations of physicians. Although I could rise in outrage at the idea of the concept of justice at the bedside in 1981, less than twenty years later it has a secure place at the head of the patient's bed to ensure that the patient gets a fair share (but not more) of the medical resources and that the social system gets its money's worth.

There have been significant changes in clinical medicine of all three of the *Belmont* principles — beneficence, respect for persons, and justice — over a period of some four decades, and especially in the last twenty years. The meaning of beneficence has shifted from acting for the good of the sick person to acting for the good of a body part or physiological system. Respect for persons has been redefined from concern for the sick person to almost solely as the right of the patient to choose independently from among all options. Justice was originally not seen to apply to clinical medicine; now it is apposite because of the shift in viewpoint of the medical act from that of an individual clinician for an individual patient to a commodity or a resource within a marketplace social system.

The actors in the medical drama have become atomistic individuals, their relationships devalued, and treatment has become the increasingly successful therapy of body parts or systems. Scientific, legal, and marketplace worldviews have increasingly defined the participants and their actions, with medicine reflecting the changes that have occurred in the surrounding society.

There are, of course, countervailing forces in which the patient rather than the disease is the object of medicine. I have already written extensively about this perspective, which has been developing since the late 1920s. It would be incorrect, despite its demonstrable advantages to patients and the social system, to portray it as more than an alternative viewpoint at this time. Putting the patient at the center has a nostalgic quality, but in this newer incarnation, the sick or well person is not merely the moral center of medicine, but the logical and intellectual center. Nostalgia alone is an inadequate argument.

A final case illustrates the present trend of treating individual body parts rather than individual persons. A forty-nine-year-old woman developed recurrent breast cancer three years after a lumpectomy, radiation, and chemotherapy. It progressed very rapidly, so that within a few weeks the cancer had spread to the lungs, bones, and liver. The severity of her liver disease made adequate chemotherapy impossible, but her oncologist continued to talk of cure "once the liver is better." When she became sicker and deeply jaundiced, she was admitted to a major teaching hospital. Because of gross edema and abnormalities of electrolytes, a nephrologist was called, who took over the problem of kidney function. Her liver function worsened, but the oncologist's stated optimism did not wane. The residents were kind and attentive, but busied themselves with her abnormal liver function. She and her partner, supported by the physicians, continued to make plans for her future and would not hear of the possibility that she might die. Reluctantly, she accepted the advice that her parents be told of her illness. She was discharged from the hospital but was readmitted in three days with a pathologic fracture of the hip. The hip was pinned, but postoperatively her liver function worsened and her blood pressure fell. She was transferred to an intensive care unit. The oncologist said that as soon as the problems with her liver and kidneys were straightened out, he could start treating her cancer. In a few days, the nephrologist announced that her kidneys were now doing well. She became more sick and confused. The orthopedist came and pronounced the wound healing well. He asked the nurses whether they could get her up and walking. She died the next morning.

Notes

1. "The Principles of the *Belmont Report* Revisited: How Have Respect for Persons, Beneficence, and Justice Been Applied to Clinical Medicine?" *Hastings Center Report* 30 (2000): 12–21.

2. A contrary example is the rehabilitation medicine movement brought into being by the residua of World War II wounds and infirmities, which brought new understandings of function, thereafter defined not solely by the action of a body part but by the ability of a person to participate in a social role. Rehabilitation may not correct the underlying pathogenic mechanism, but it can restore function by concentrating on teaching the person to compensate for lost function and teaching accommodation to impairments. Rehabilitation medicine, despite

its good works, remains peripheral to mainstream medicine, not part of medicine's ego idea.

3. Eric J. Cassell, "Do Justice, Love Mercy: The Inappropriateness of the Concept of Justice Applied to Bedside Decisions," in *Justice and Health Care*, ed. Earl E. Sharp (Dordrecht/Boston/Lancaster: D. Reidel Publishing Company, 1981).

4. Jon Elster, *Local Justice: How Institutions Allocate Scarce Goods and Necessary Burdens* (New York: Russell Sage Foundation, 1992).

5. Ibid., 146–48.

6. Lynn A. Jansen, "Local Justice and Health Care: The Case of Pain Management" (paper presented at a medical ethics meeting at Johns Hopkins University, Baltimore, Maryland, 1999).

PART III

The *Belmont* Principles: Possibilities, Limitations, and Unresolved Questions

7 We Sure Are Older But Are We Wiser?

Karen Lebacqz

As one who was privileged to serve on the National Commission that developed the *Belmont Report*, I have watched this report's history of use with some interest. I begin this chapter with an observation about the use and interpretation of the principles articulated in the *Belmont Report*, and then turn to a related and more crucial question: Did the *Belmont Report* give rise to what is now called, somewhat pejoratively, "principlism," and should we eschew the principles? With particular attention to feminist critiques, I will argue for a retrieval of the principles of the *Belmont Report* and for their continued interpretation and better use.

Interpretation and Use of the Principle of Respect for Persons

The *Belmont Report* identified three principles: respect for persons, beneficence, and justice.[1] These terms are somewhat vague. Although the National Commission gave preliminary definition to each of them, the vagueness of the terms has allowed for refinement and reinterpretation over time. In my view, some of that reinterpretation has diminished rather than enhanced the original meanings. I will illustrate this contention by looking at the principle of *respect for persons*.

As originally specified by the National Commission, "respect for persons incorporates at least two basic ethical convictions: first, that individuals should be treated as autonomous agents, and second, that persons with diminished autonomy are entitled to protection."[2] In short, the National Commission identified two crucial aspects of the principle of respect for persons. To respect an autonomous person, one "gives weight to that person's considered opinions and choices while refraining from obstructing their actions" unless those actions are clearly detrimental to others.[3] To respect a nonautonomous person, one provides appropriate protection for that person. The requirement to respect nonautonomous persons does not disappear; it simply takes a different form than does the requirement to respect autonomous persons. In addition, the National Commission left open the possibility that there would be additional important "convictions" covered by this fundamental principle.

Two things have happened in subsequent history, however, that truncate and diminish the power of this fundamental principle. First, respect for persons became interpreted solely in the language of respect for autonomy. Second, autonomy itself became interpreted in restricted ways.

RESPECT FOR PERSONS BECOMES RESPECT FOR AUTONOMY

In every edition of their best-selling textbook *Principles of Biomedical Ethics*,[4] Beauchamp and Childress use the language of "autonomy"—and subsequently "respect for autonomy"—rather than "respect for persons."[5] This leads them to say that our obligations to respect autonomy do not extend to nonautonomous persons.[6]

This truncating of respect for persons into respect for autonomy is unfortunate in two ways. First, the respect due to nonautonomous persons is lost. In the National Commission's view, *all* people, *all* human beings, are deserving of respect. The specific form of that respect will differ depending on the stage of life and the capabilities of the person (e.g., whether or not she or he has autonomous agency), but there is no diminishment of a need to respect those who lack autonomy. While Beauchamp and Childress intend to provide protections for nonautonomous persons under the principle of nonmaleficence, the shift in language suggests that nonautonomous persons are not deserving of respect; this is a crucial and unfortunate move. The early blastocyst, for example, is not autonomous; under a principle of respect for autonomy, it would not be deserving of respect. Yet the ethical perception of many people is that human life is deserving of respect at all stages, not simply when autonomy is present. For example, the Ethics Advisory Board of Geron Corporation has declared flatly that even the early blastocyst is deserving of respect.[7]

The second unfortunate effect of the truncating of respect for persons into respect for autonomy is the exclusive focus on *autonomy* in defining persons. Subsequent feminist literature has stressed not autonomy and its implications of choice and freedom, but *relationality* and its implications of connection and commitment. The National Commission may be partly to blame for the focus on autonomy: the *Belmont Report* appears to locate the core of respect for persons quite solidly in the question of autonomy. However, the National Commission also claimed that the principle of respect for persons incorporates *at least* basic convictions about autonomy; however, it left room for additional understandings of the importance of respect for persons that might extend beyond respect for autonomy. This can become crucial, as I will show in my revision of the principle in the following section.

AUTONOMY AS FREEDOM OF CHOICE

The reduction of respect for persons to respect for autonomy is not the only problem in the history of interpretation of this principle. Not only was respect for persons reduced to respect for autonomy, but "autonomy" was also reduced. Autonomy has been interpreted along the lines of John Stuart Mill's "liberty" rather than along the lines of true Kantian "autonomy." In the President's Commission report *Making Health Care Decisions*, for example, the phrase was changed to "respecting self-determination," and self-determination was defined as "an individual's exercise of the capacity to form, revise, and pursue personal plans for life."[8] That is, autonomy (which the President's Commission appears to equate with self-determination) is interpreted in terms of allowing individual choice of whatever people want for themselves. Any Kantian understanding that each choice must be implicitly universalized is completely absent.[9] Thus, in our individualistic culture, "autonomy" has lost its universal law-giving connotation and has become merely a question of being left alone to choose what one wishes.

Here again, the National Commission's language no doubt participates in and contributes to this interpretation. In a statement reflecting Mill's rejection of paternalism, the National Commission declared, "To respect autonomy is to give weight to autonomous persons' considered opinions and choices while refraining from obstructing their actions unless they are clearly detrimental to others."[10] Unfortunately, the net result of such an interpretation is that respect for persons becomes reduced to unfettered freedom of choice. Such an understanding is exemplified in John Robertson's argument that couples should be free to choose to have children or not, to choose the characteristics of their children, and to use any and all available technologies in order to effect either of the first two choices.[11]

As respect for persons is reduced to autonomy and autonomy to self-determination or freedom of choice, the logical outcome is that the broad-ranging *principle* of "respect for persons" is then truncated into the *rule* of "informed consent." The only question we ask is whether one *consented* to medical interventions or use of technologies. Any question as to whether such interventions or technologies are right is utterly lost. In fact, however, requiring informed consent is only one possible interpretation of the principle of respect for persons.

Reclaiming the Broader Context and Intent

As noted earlier, the National Commission may have given some impetus to this developmental line and hence must take some responsibility for the truncating of respect for persons. Nonetheless, the development of the principles approach to bioethics over the last twenty years cannot be completely laid at the National Commission's door. In its discussion of informed consent as one interpretation of autonomy, the National Commission pointed carefully to the fact that certain populations (e.g., prisoners) can live under conditions that compromise autonomy, and hence a simple interpretation of respect for persons as necessitating freedom of choice is not adequate.[12] In subsequent development, the National Commission's caution was not heeded.

In short, the truncating of respect for persons into respect for autonomy has had several unfortunate effects that severely limit the original intent of the National Commission as I recall it. The broader principle of respect for persons, as enunciated by the National Commission, may take primary form in respect for autonomy, but it takes secondary form in protection of those who are not autonomous, and it may have other implications as well. Nor is it clear that autonomy alone is sufficient to define what respect for persons would require. There may be other aspects of persons, such as their embeddedness in historical context, that are also fully deserving of respect. This alone should signal that "respect for persons" as a principle would require a much more nuanced and refined understanding of what is required than is sometimes found in the literature. For example, to respect persons, rather than simply respecting autonomy, may be to respect the social context in which persons gain and express their identity. This social context is traditionally part of Roman Catholic understanding and is also emphasized in feminist theory.

In Roman Catholic tradition, the person is first and foremost social. Individual human flourishing is never separate from the common good. One does not develop the virtues, or "excellences" of human life, simply by making autonomous choices. Participation in social life is not simply a right, it is a duty. In an essay that significantly influenced some members of the National Commission, Richard McCormick argued that a child who lacked the ability to give autonomous consent might nonetheless be used in research involving no direct benefit to the child, because learning to participate in fostering the common good is part and parcel of becoming the kind of person the child should become.[13] This insight can be paraphrased as "the child would consent if the child could consent because the child should consent." It is precisely the sense of "should" or obligation that is utterly missing when respect for persons is truncated into freedom of individual choice based on preferences and desires.[14]

Similarly, feminists stress the social context of human life. In her critique of the principles approach to bioethics, Christine Gudorf notes that feminists would complement the standard account of principles with additional principles that are more social in nature: mutuality, community, solidarity, empathy, and the like.[15] Many feminists have argued that human life is rarely about autonomous choice. Nancy Hirschmann, for example, suggests that we are born into relationships that give us obligations that are not chosen. Thus, an ethics that begins with autonomy in the sense of free choice does not do justice to the complexity of human relationships or to the features of life that often obligate us.[16] Respect for persons, in such a view, would involve not simply honoring freedom of choice but respecting the kinds of obligations and relationships in which people find themselves in the course of living. "Care" supplements "justice" as part of the mantra of ethical living.

In short, feminists, along with others, would urge a wider range of ethical principles than they find in the tradition reflected in the *Belmont Report*. I believe that this search for wider principles reflects in part the fact that the interpretive tradition has truncated the original principles. Although the National Commission is not totally responsible for the truncating of its vision and principles, it may still be fair to ask whether the commission did enough to protect against the likely trend in interpretation and use of those principles within an individualistic, liberal culture that so touts individual freedom of choice. The National Commission in its work offered no fundamental critique of liberal culture and hence perhaps failed to protect its own vision from the stultifying effects of contemporary liberal society.

Are Principles Bankrupt?

The misuse and limited interpretation of principles is not the only significant issue that can be raised about the National Commission's work and its legacy. Is the dependence on principles itself bankrupt? Perhaps the problem lies not with particular principles, but with the use of principles in general. If so, perhaps we should henceforth eschew principles, and throw out the *Belmont Report* altogether, seeing it as of historical interest but of little lasting value.

Principles have come under attack from a number of quarters, and many of those attacks are gathered in the volume edited by DuBose, Hamel, and O'Connell titled *A Matter of Principles: Ferment in U.S. Bioethics*.[17] I have already indicated that Gudorf, representing feminist views in that volume, brings a rather sharp critique of the specific principles enshrined by Beauchamp and Childress. However, feminist critique goes beyond a simple rejection of particular principles.

As Gudorf notes, "the feminist analysis . . . maintains that principlism, because of its reliance on principles and a specific set of principles, functions to create and sustain a unique social system."[18] In other words, the very dependence on principles, as well as the particular principles chosen, not only reflects a "male-centered" view but serves to sustain and strengthen that view — to the detriment of women. Other feminists also suggest that feminist bioethics needs to "rethink fundamental assumptions," including not only the substance of the principles but "philosophical methods" themselves.[19]

These feminist critiques border on a total rejection of principles (one is tempted to say, "in principle"!). At stake is the understanding that, no matter the specific content of the principles, the very use of principles as a method ignores the "concrete other" and evades the important nuancing that must enter every adequate ethical judgment. Drawing on the work of Seyla Benhabib[20] and Martha Nussbaum,[21] for example, feminists would urge that the use of principles tends to make ethical decisions into abstract formulations that will always violate the particularity of concrete subjects. In Aristotelian or "communicative" style, contemporary feminists would tend to urge instead an ethical perspective that begins with the concrete, lived, and painful experiences of women and moves from those experiences toward advocacy and action, rather than beginning with an abstract, predetermined set of principles and drawing conclusions for "cases" from those principles.[22]

At first glance, then, it might seem that a feminist approach to bioethics would simply eschew principles. Yet I believe that the issue is not so simple. As Gudorf notes, feminists increasingly reject any dichotomy between an ethic of concrete "care" and the use of principles such as justice.[23] Similarly, Susan Sherwin, whose *No Longer Patient* has become a standard in feminist bioethics, notes that "many feminist ethicists still envision a place for principles in ethics" and that her own opposition to oppression "is a principled one."[24]

Some principles would appear to be needed in bioethics, in order to serve the following functions:

1. If there is ever to be cross-cultural critique or analysis (e.g., of the practice of cutting women's genitals), some principles will be necessary. All feminists believe that oppression is wrong, and therefore I would suggest that a principle of justice that rejects oppression is implicitly recognized by feminists, even by those who eschew principles.

2. While no one would propose that principles are the *whole* of bioethics, principles are necessary to social policy. Guidelines are needed to set policy for predictive genetic testing, for example, and those guidelines will implicitly or explicitly reflect and embody principles.

3. The *Belmont Report* explicitly notes that rules are often inadequate and suggests that "broader ethical principles will provide a basis on which specific rules may be formulated, criticized, and interpreted."[25] Some of the opposition to principles arises because principles get turned into *rules* rather than kept as principles that allow space for the critique and interpretation of rules. Surely some of the feminist opposition to principles, for example, is based on an understanding that a simple rule requiring "informed consent" may ignore the contextual realities for women whose "consent" may be coerced (e.g., in cases of in vitro fertilization). But opposition to the deadening application of a principle such as respect for persons does not mean that feminists would support *dis*-respect for women. The fundamental principle still holds; what is at stake in the opposition to principles is the way in which principles are understood and interpreted. Thus, some of the opposition is to what is called "principlism," not to principles per se. I am therefore in large agreement with Beauchamp and Childress when they claim that "only a faulty conception of the nature and interpretation of principles would lead to the conclusion that principles have no integral role in moral reasoning in concrete circumstances."[26]

Retrieving and Reclaiming the Principles

If there is a place — even a necessity — for principles, then how would we assess the validity of the original principles proposed by the National Commission in the *Belmont Report*? Did the National Commission err in its selection

or exposition of principles? The National Commission identified three principles, but noted that "other principles may also be relevant."[27] The three
principles were understood to be among those "generally accepted in our cultural tradition."[28] To some extent, the National Commission was constrained
by having to choose principles that could be affirmed broadly in a pluralistic
culture. Are those principles adequate or are additional or alternative principles needed, as Gudorf suggests? If I were to choose principles today, would
I choose the same ones, or different ones?

1. RESPECT FOR PERSONS AND COMMUNITIES

With the increasing significance of genetics, it becomes particularly crucial
that we find ways of respecting people that acknowledge families. With
increasing globalization, it becomes especially important not to impose an
atomistic concept of persons on people from cultures that have strong community identity. So I favor the retention of a principle of respect for persons
but suggest that such a principle must be interpreted within a more communal understanding and tradition than has permeated the informed-consent
approach. We need a broad understanding of the fundamental relationality or
sociality of persons. Whether we take such an understanding from classical
Roman Catholic notions of the common good, from Protestant commitments to covenant, from philosophical understandings of the narrative and
embedded nature of human life, or from the conviction of George Herbert
Mead and subsequent social psychology that we are formed only in relationship, human beings are not simply autonomous individuals. We are born into
relationships and would not grow without them. We become who we are in
community. Hence, the principle of respect for persons should include
respect, not simply for individual autonomy but for relationality — for family
members and even for social units and cultural traditions.

If I were to modify this principle today, I would draw on the Western tradition of *covenant*.[29] As a concept, covenant implies a deeper connection of
the individual with the community. It places researcher and subject in colle-
gial relationship. It implies that what happens to one affects all. Recognition
of covenant would enrich respect for persons by making clear that persons
are not simply atomistic individuals but are members of communities, both
large and small, and that respect for persons must include respect for those
communities and their traditions. Recognition of covenant would make clear
that actions affecting one person also affect others and that we are bound
together in multiple ways that undermine any easy freedom of choice.

2. BENEFICENCE AND NONMALEFICENCE

In its work, the National Commission lumped together beneficence and nonmaleficence. Whether joined or separated, I believe that each retains an
important place in bioethics.

The mandate to do good — to find cures for devastating diseases, to
understand growth and disease processes with more accuracy so as to refine

interventions, and so on — underlies medical and scientific research. More-over, doing good for individual subjects, as well as being concerned about doing good overall, seems imperative. Particularly for subjects in nontherapeutic research or in double-blind trials who may not be receiving appropriate interventions and whose treatment must be carefully regulated for purposes of contributing to research, it is very important to pay special attention to times when a subject should be pulled off a trial because of individual failure to thrive. Thus, the principle of beneficence and its counterpart non-maleficence still has important implications in research ethics.

Still, if I were to write new principles today, I would draw on the religious traditions that underlie American society and invoke a concern for *compassion* to supplement beneficence. To have compassion is to suffer with another, not simply to do good for another. Compassion first requires that we receive the suffering of the other, taking part of the reality of the other into ourselves. It therefore serves as a caution — and, I believe on some level, a protection — against the kind of paternalism and imperialism that assumes we know what is good for others.[30] Only those who suffer know what would alleviate that suffering, and often what is needed is not cures or new therapies, but rather community and accompaniment.

Often, too, what is needed is financial support and social involvement. Roman Catholic tradition here invokes a principle of participation.[31] It is a duty of human beings to participate in society and to contribute to the common good. But precisely because it is such a duty, it is concomitantly a duty of society to enable and facilitate that participation.[32] Compassion means listening to the voices of those who are oppressed and allowing research agendas to be determined by what those who suffer need for their empowerment. Compassion moves beyond doing good *to* or *for* others and into working *with* them to effect a more just and life-supporting world.

3. JUSTICE AND LIBERATION

The mention of empowerment brings me to a consideration of justice. The *Belmont Report* says that "an injustice occurs when some benefit to which a person is entitled is denied without good reason or when some burden is imposed unduly."[33] The National Commission also adopted the classical philosophical formulation that "equals ought to be treated equally."[34] The National Commission interpreted justice questions somewhat narrowly in the research arena, limiting the discussion largely to the question of who should serve as subjects of research. This is surely an important question. One of the topics of feminist critiques has been that women are not included as research subjects in a great deal of research, which leaves us with serious lacunae in our knowledge of disease processes and treatments for women.[35]

But is this focus on selection of subjects enough to ensure justice in research? I think not. Since the time of the National Commission's work, justice has been the subject of a number of important philosophical treatises: Michael Walzer's *Spheres of Justice*,[36] Bruce Ackerman's *Social Justice in the*

Liberal State,[37] John Rawls's revised version of his theory of justice offered in *Political Liberalism,*[38] and perhaps most significantly, Iris Marion Young's *Justice and the Politics of Difference.*[39] These treatises go not to micro-questions of justice, such as who should shoulder particular burdens of research, but to macro-questions: Do liberal philosophy and politics serve justice? How is fairness of social organization to be judged across cultures? Social justice in this broader sense is a particular passion of mine, and I have done some writing on it[40] — so I would retain the term "justice." However, I think the term needs more refinement than it received from the National Commission.

Here again I believe that there is an understanding rooted in several of the religious traditions that underlie this country that would broaden and extend our understanding of what justice in research might require. In a pluralistic society, it may seem suspect to draw on faith convictions for the interpretation of a general term such as "justice." However, it is not possible to avoid faith convictions: those who interpret justice in Platonic or Aristotelian terms are also drawing on particular traditions that have their own limitations and are culturally specific. In this sense, they too depend on faith convictions. In the background of American culture, there is a stream of thought that allows an interpretation of justice that goes beyond simply stating that equals should be treated equally.

Justice may also be understood as requiring *attention to power differentials and to the liberation of the oppressed.* In Judeo-Christian religious tradition, this theme is prominent: when the prophet Amos thunders that justice will roll like a mighty stream (Amos 5:24), he is talking about the kind of understanding of justice that redresses the wrongs done by those in power. Justice in ancient Israelite tradition and in subsequent Christian tradition always attended to the ways in which the poor and oppressed were beaten down by the system, the ways in which rulers contributed to the disempowerment of the poor ("trampling" on them, as Amos puts it — 5:11) and the ways in which the poor are to be liberated and "lifted up" and the rich sent "empty away" (Luke 1:52–53). Justice is not simply about treating equals equally. It is about equalizing things that are not equal. It is about redressing imbalances of power. While this theme has been particularly prominent in religious tradition, it has been picked up in feminist philosophy as well (see, e.g., Iris Marion Young's *Justice and the Politics of Difference*). Further, it has strong roots in the founding of the United States as a country dedicated to liberation of those who had experienced political and social oppression.

Such a broader understanding of justice would require looking not simply at the selection of research subjects but at entire systems and structures and how they distribute power and privilege. Doing so might require some fundamental critique of the research enterprise itself — of the forms of internal review that determine which projects get funded, of the ways in which universities are making links with biotechnical and pharmaceutical companies that shape the kinds of projects that will receive priority and support, and so on. The National Commission did not raise these kinds of questions. We accepted the research enterprise as it stood in the United States at that time. We asked questions *within* that framework, but did not question the basic framework itself. Such questions have yet to be raised with seriousness in the dominant bioethics community, though they have been the meat of feminist bioethics for some time.

Conclusions

In sum, I would contend that

- the principles enunciated in the *Belmont Report* have been misinterpreted in the twenty years since the National Commission's work and that such misinterpretations have led to the development of an unfortunate kind of principlism that is rightly criticized;
- there is a need for principles in the shaping of public policy in the research arena; and
- the principles enunciated by the National Commission still retain validity.

However, the National Commission's principles are in need of extension and reinterpretation, both to retrieve their original depth and breadth and also to attend to subsequent developments in the refinement of understandings of oppression and liberation.

Notes

1. The National Commission for the Protection of Human Subjects of Biomedical and Behavioral Research, *The Belmont Report: Ethical Principles and Guidelines for the Protection of Human Subjects of Research* (Washington, DC: U.S. Government Printing Office [DHEW Publication # (OS) 78-0012], 1978).

2. National Commission, *Belmont Report*, 4.

3. National Commission, *Belmont Report*, 5.

4. Tom L. Beauchamp and James F. Childress, *Principles of Biomedical Ethics* (New York: Oxford University Press, 1979).

5. In the first several editions they call it the principle of autonomy; this is later (3rd edition) corrected to be the principle of *respect for autonomy*. At no time, however, does the wider language of respect for persons frame the discussion.

6. Beauchamp and Childress, *Principles of Biomedical Ethics*, 65, 127.

7. Geron Ethics Advisory Board, "Research with Human Embryonic Stem Cells: Ethical Considerations," *Hastings Center Report* 29, no. 2 (1999): 32.

8. President's Commission for the Study of Ethical Problems in Medicine and Biomedical and Behavioral Research, *Making Health Care Decisions*, vol. 1 (Washington, DC: U.S. Government Printing Office, 1982), 44.

9. The President's Commission suggests that self-determination is "both a shield and a sword." See *Making Health Care Decisions*, 45. As a shield, it protects

the person from outside control. As a sword, it allows individuals to define "their own particular values" (p. 46). Here, autonomy as self-determination clearly reflects John Stuart Mill's rejection of paternalism, but at the same time is reduced to "personal integration within a chosen life-style" (p. 46).

10. National Commission, *Belmont Report*, 5.

11. John A. Robertson, *Children of Choice: Freedom and the New Reproductive Technologies* (Princeton, NJ: Princeton University Press, 1994).

12. National Commission, *Belmont Report*, 6.

13. Richard A. McCormick, "Proxy Consent in the Experimentation Situation," *Perspectives in Biology and Medicine* 18 (1974): 2–20.

14. For a thoroughgoing critique of the liberty or rights approach, and one that emphasizes duty and obligation instead, see Benjamin Freedman, *Duty and Healing* (New York: Routledge, 1999).

15. Christine E. Gudorf, "A Feminist Critique of Biomedical Principlism," in *A Matter of Principles: Ferment in U.S. Bioethics*, ed. Edwin R. Dubose, Ron Hamel, and Laurence J. O'Connell, 167 (Valley Forge, PA: Trinity Press International, 1994).

16. Nancy J. Hirschmann, "Freedom, Recognition, and Obligation: A Feminist Approach to Political Theory," *American Political Science Review* 83, no. 4 (1989): 1227–44.

17. Edwin R. Dubose, Ron Hamel, and Laurence J. O'Connell, eds., *A Matter of Principles: Ferment in U.S. Bioethics* (Valley Forge, PA: Trinity Press International, 1994).

18. Gudorf, "A Feminist Critique," 165.

19. Laura M. Purdy, "A Call to Heal Ethics," in *Feminist Perspectives in Medical Ethics*, ed. Helen Bequaert Holmes and Laura M. Purdy, 12 (Bloomington: Indiana University Press, 1992).

20. Seyla Benhabib, *Situating the Self: Gender, Community and Postmodernism in Contemporary Ethics* (New York: Routledge, 1992).

21. Martha C. Nussbaum, *Love's Knowledge: Essays on Philosophy and Literature* (New York: Oxford University Press, 1990).

22. See Karen Lebacqz, "Feminism and Bioethics," *Second Opinion* 17, no. 2 (1991): 11–25.

23. Susan Sherwin, *No Longer Patient: Feminist Ethics and Health Care* (Philadelphia, PA: Temple University Press, 1992), 82.

25. National Commission, *Belmont Report*, 1.

26. Beauchamp and Childress, *Principles of Biomedical Ethics*, 4th ed., 107.

27. National Commission, *Belmont Report*, 2.

28. Ibid., 4.

29. Cf. William F. May, *The Physician's Covenant: Images of the Healer in Medical Ethics* (Philadelphia, PA: Westminster Press, 1983).

30. Karen Lebacqz, "The Weeping Womb: Why Beneficence Needs the Still, Small Voice of Compassion," *Secular Bioethics in Theological Perspective*, ed. Earl Shelp, 85–96 (Dordrecht, Neth.: Kluwer Academic Publishers, 1996).

31. Pope John XXIII, *Pacem in Terris*, April 11, 1963, paragraph 26.

32. Ten years ago, the Americans with Disabilities Act (ADA) was passed. Its purpose was to ensure that people with disabilities would not be discriminated against in employment opportunities. Today, the evidence is overwhelming that the ADA, far from securing nondiscrimination for people with disabilities, has probably contributed to their marginalization. We have failed to provide for participation. Official policy on disability is still beneficence based but fails to listen to the voices of disabled people. See, e.g., Harlan Lane, *The Mask of Benevolence: Disabling the Deaf Community* (New York: Random House, 1993).

33. National Commission, *Belmont Report*, 8.

34. Ibid.

35. See, e.g., Leslie Laurence and Beth Weinhouse, *Outrageous Practices: How Gender Bias Threatens Women's Health* (New Brunswick, NJ: Rutgers University Press, 1994), chapter 3.

36. Michael Walzer, *Spheres of Justice: A Defense of Pluralism and Equality* (New York: Basic Books, 1983).

37. Bruce A. Ackerman, *Social Justice in the Liberal State* (New Haven, CT: Yale University Press, 1980).

38. John Rawls, *Political Liberalism* (New York: Columbia University Press, 1993). At the time of the National Commission's work, Rawls's *A Theory of Justice* (Cambridge, MA: Harvard University Press, 1971) had just been published.

39. Iris Marion Young, *Justice and the Politics of Difference* (Princeton, NJ: Princeton University Press, 1990).

40. Karen Lebacqz, *Six Theories of Justice* (Minneapolis, MN: Augsburg, 1986); *Justice in an Unjust World* (Minneapolis, MN: Augsburg, 1987).

8 Toward a More Robust Autonomy

Revising the *Belmont Report*

Larry R. Churchill

The flaw in the *Belmont Report* lies less in its principles and more in its interpretation of one of those principles, indeed the key principle, respect for persons. Rather than a strong endorsement of respect for the autonomous choices of subjects, *Belmont*'s understanding of respect for persons presents a weak and distorted understanding of self-determination. My main purpose here is to show how this flawed notion of autonomy in *Belmont* can be damaging to the enterprise of research with human subjects and to suggest an alternative formulation.

Respect for Persons, Disrespect for Autonomy

In the *Belmont Report*, respect for persons is listed as the first of "three basic principles, among those accepted in our cultural tradition" that are "particularly relevant to the ethics of research involving human subjects."[1] The two remaining principles are beneficence and justice. Respect for persons is interpreted as incorporating two distinct convictions: "First, that individuals should be treated as autonomous agents, and second, that persons with diminished autonomy are entitled to protection."[2] *Belmont*'s interpretation of what it means to respect autonomy reads as follows: "To respect autonomy is to give weight to autonomous persons' considered opinions and choices while refraining from obstructing their actions unless they are clearly detrimental to others."[3] This explication of respect for autonomy is inadequate and troubling in three ways.

First, *Belmont*'s stress on "giving weight" is anemic in comparison to the usual understanding of respect for autonomy as recognizing that individuals have final authority for decisions or that they are free and self-determining moral agents. For someone to "give weight" to a research subject's opinions and choices implies that the authority to weigh and judge resides somewhere other than with the subject, presumably with the investigator. A relationship is implied, but its nature is left unclear, except that the research subject is given a passive role. Thus *Belmont* shifts moral agency away from the subject. Moreover, the metaphor of weighing raises questions about what counterbalances will be placed on the scales and what authority or "weight" these counterbalances will be given. In other words, investigators are encouraged to demonstrate respect for a research subject's opinions and choices, but are not required to give those opinions and choices a primary and preemptive status in the informed consent process. Accordingly, the investigator becomes the chief moral actor with authority over the deliberations of how and to what extent to follow the subject's opinions and choices. "Weighing" thus serves in this context as a metaphor of paternalism and bespeaks an eviscerated notion of autonomy. Autonomy, as *Belmont* sees it, is not a principle, but one of two subheadings under a principle, with far more space devoted to explication of the obligation to respect persons by protecting those with diminished capacity than to the meaning of autonomy.

Second, it is unclear why respect is restricted to individuals' "considered opinions." What of their *un*considered or *ill*-considered opinions? Here we have a formulation employing a notion of autonomy derived from therapeutic contexts, which when applied to research contexts is inadequate and misleading. Persons who are invited into research should be free to decline participation for any and all reasons — considered, well-considered, ill-considered, or completely unconsidered. No one should place conditions on the refusal to enter a research protocol, although in some circumstances it does make sense to place conditions on assent to enter a protocol, for example, when subjects might enroll with unfounded hopes of a cure, or when the risks are disproportionate to the potential for benefit.

A third inadequacy of *Belmont*'s interpretation of autonomy follows from the first two. If a subject's "considered" opinions and choices are to be "respected" only in the sense of being given some weight, there is no strong obligation on the part of researchers to promote independent choices or enable subjects to make the kind of knowledgeable choices that actually reflect their values, based on an adequate understanding of what they are choosing. Seeking the knowledgeable choices of subjects does not seem to be the meaning of autonomy or the goal of informed consent, as *Belmont* interprets it under respect for persons. It should be.

Promoting Self-Determination in Subjects

Beauchamp and Childress describe a more adequate notion of this principle, which they call respect for autonomy, in *Principles of Biomedical Ethics*. Their principle "involves treating persons to enable them to act autonomously."[4] They further state that "discharging the obligation to respect patients' autonomy requires equipping them to overcome their sense of dependence and achieve as much control as possible or as they desire," and they add, "respecting another includes efforts to foster and effect that person's outlook on his or her interests."[5] Note here the active agency of the physician in promoting and enabling independent judgment in the patient rather than in assuming the role of arbiter of the other person's opinions and values, as *Belmont* presents it.

Beauchamp and Childress clearly have in mind a physician-patient relationship as the context for their views. If their ideas provide a standard for patient care, this high standard should also apply to research with human subjects. If ordinary clinical decision making must be patient-centered in the way they describe, then decision making about entry into a research project must be at least this ambitious in promoting and honoring free and independent judgments in subjects. Indeed, I believe that something similar to the Beauchamp and Childress formulation is widely (and appropriately) endorsed by bioethicists and physicians as the standard for both treatment and research. Yet such an emphasis is missing from *Belmont*, and in its place is a weakened role for subjects and a controlling role for investigators.

Any adequate principle of autonomy begins with the conviction that competent patients and subjects *already* possess the right to decide for themselves. The deliberative process is lodged with subjects. Even though a robust deliberative process can only be achieved with the help of investigators, investigators do not have decisional prerogatives. These belong to subjects, and only to subjects, so long as they are adults and possess decisional capacity. The investigator's role is a critical one, for it involves information giving and initiation of a reflective process that will strengthen subjects' decisional abilities. This enables subjects to bring to decisions about participation not only reliable information about the research protocol but also the personal life values that are germane to their choices. Thus the investigator's role is an essential one, but it is not the decisive one.

The chief point here is that investigators should not be viewed as morally empowered to "give weight" to subjects' choices. Subjects are not given authority for self-determination by investigators. Subjects possess this authority because they are persons. This is what respect for persons must mean if it is to entail respect for autonomy. The appropriate action for investigators is, therefore, not to give authority to subjects' values, but to acknowledge subjects' right to independent decisional authority based on their own values. Moreover, investigators are obligated not only to acknowledge the right of independent choice but to *promote*, to the extent possible, that right in subjects for whom deliberation about choices may be compromised by illness, uncertainty, fear, or lack of information. To do less is to fail to respect them as persons, by failing to help them "overcome dependence" or "foster and effect [the subject's] outlook on his or her interests," as Beauchamp and Childress have aptly put it.

Further evidence of the inadequacy of *Belmont*'s understanding of these issues is provided in the examples used to illustrate the negative case, or the lack of respect for an autonomous agent. "To show lack of respect for an autonomous agent," *Belmont* claims, "is to repudiate that person's considered judgments, to deny an individual the freedom to act on those considered judgments, or to withhold information necessary to make a considered judgment, when there are no compelling reasons to do so."[6] *Belmont* does not specify what might qualify as a compelling reason to withhold information necessary for a research subject to make a considered judgment, but this qualification carries the paternalistic tone associated with withholding information for the patient's own good in therapeutic contexts.

My main concern with *Belmont*'s examples of lack of respect is their extreme nature. "Repudiating," "denying," and "withholding" are such egregious examples of lack of respect for autonomy that it leaves the reader unclear about the more probable and more subtle actions that fall short of these obviously unethical examples. For example, what about simply allowing subjects to be more optimistic about the potential for a direct, therapeutic effect from research participation than the evidence warrants, or failing to separate clearly the differing goals and burdens imposed between research protocols and other options available to a patient? I am not suggesting that *Belmont*, which is necessarily a brief document, should discuss the great variety of more subtle cases in which the autonomy of subjects can be compromised. I am suggesting, however, that when taken together with its weak interpretation of autonomy, these few and extreme examples do nothing to suggest that a strong autonomy principle is important. It is quite possible for investigators to believe that if they do not "repudiate," "deny," or "withhold," then their actions are consistent with respecting autonomy.

Objections

Some may argue that this proactive enablement of patients is an excessively burdensome demand on physicians in ordinary therapeutic contexts. I would disagree. Yet whatever the standard for ordinary practice, it is incumbent upon investigators to enable their subjects as part of obtaining a valid consent precisely because the research alliance between investigators and subjects is one in which therapeutic results are secondary to the goals of hypothesis testing and generalizable knowledge. This means that research is an activity that does not warrant deciding for others or withholding information for the other person's own good, so long as they are decisionally capable. Paternalistic actions and attitudes have no place in the decisions about whether persons will become subjects and enter into research protocols. Yet it is something dangerously akin to such actions and attitudes that are identified with the autonomy principle in *Belmont*.

Proactive enablement as a duty assigned to investigators might evoke another sort of criticism—that it is too idealistic and can rarely be achieved. This objection seems weak. Why foreclose on the possibility of a deeper and more meaningful autonomy because some subjects may not be capable of it? Some subjects—perhaps the majority—are undoubtedly capable and desirous, and unless researchers aim for a more meaningful exercise of autonomy, these subjects will be stuck in a passive role. The key to achieving a more active and authentic role for subjects may lie in the investigator's determination to move beyond the idea of consent as a Miranda warning about potential risks and benefits and toward a more robust and subject-centered idea of informed decision making.

The duty to enable is not idealist, but realistic, for it recognizes the asymmetry of power between investigator and subject. More importantly, a duty to enable responds to the fundamentally social nature of the exercise of autonomy. Being free to make one's own choices is not, after all, a steady capacity with which individuals are born, but a fragile social achievement. All of us must be helped by family and friends if we are to be autonomous persons. We become autonomous through the enabling actions and support of others. In the research context, this is especially important, and it means that subjects must be empowered through their relationship with investigators, that is, through investigators' willingness to discuss, disclose, and reflect with subjects on the substantive matters of research participation, including the subjects' values. The absence of an enabling duty for researchers bespeaks a view of persons that is unrealistic because it is too atomistic. It fails to recognize the crucial place of social supports, especially the support of the researcher for the subject, in making choices that are authentic and genuine. Thus, a strong, enabling dimension to the autonomy principle is required, not only because of the vulnerability of subjects, but in order to respect the social nature of autonomy.

Some may be congenial to a duty of enablement but still object that a strong emphasis on autonomy is inappropriate because they believe that the fundamentally relational nature of moral selves is distorted by a strong endorsement of individual self-determination. This objection has much to recommend it. Indeed, I am among those who have been critical of atomistic assumptions of personhood at work in bioethics in particular and in contemporary American culture more generally.[7] Yet this criticism has less weight for clinical contexts — especially for the context of clinical research — than in other settings. For example, in the area of health policy, individual autonomy has played havoc with the need to consider the well-being of populations and to honor other values besides personal choice. Likewise, although respect for autonomy is an important principle for patients in therapeutic encounters, too much stress on autonomy can lead to an isolation of the patient and distort our understanding of the way individual decisions are embedded in webs of familial values. Moreover, stressing individual autonomy to the exclusion of other values can do real harm to families. Think here of the desperately ill patient ready to try all possible remedies, even those that will greatly burden his family and the life prospects of his children for years to come. Here the failure to balance respect for autonomy with other principles warps moral understanding.

Entry into clinical trials, however, is a very different process than entry into a therapeutic relationship, and this distinction is not well understood by subjects. Subjects sometimes have unrealistic expectations of being directly benefited from participation in research, even very early research in which direct benefit is not intended.[8] Given this misconception, the hazard is not that subjects will be isolated or will harm others through their choices, but that their voices will be overwhelmed by these unfounded expectations and by the other powerful forces driving medical research. Among these additional forces are the growing commercial investments in human subjects research, the scientific and monetary ambitions of researchers and their institutions, and the career and pecuniary interests of enrollment coordinators and others involved in recruiting and maintaining subjects' participation. Given these pressures, and the possibilities for therapeutic misunderstanding, a robust sense of autonomy is essential to research ethics.

Despite their appeal, interpretations of persons as fundamentally communal or relational have their own set of hazards, especially when the role and power of communities are not critically examined. Communities can be repressive as well as supportive of the needs of the individuals that belong to them. In the past, appeals to the community, public welfare, or social good have occasionally usurped individual rights and decisional prerogatives in the name of scientific progress. Seeking a balanced approach will mean learning to appreciate the multidimensional character of persons and to stress those ethical principles that are most likely to be important in the particular context or setting in which decisions arise. For clinical research, respect for autonomy is a sine qua non.

A Conjectural History

I have argued that *Belmont* begins with a frail and compromised notion of subjects' prerogatives for independent decision making rather than a robust endorsement of autonomy. Respect for persons, as interpreted by *Belmont*, results in disrespect for autonomy, yet it seems likely that at least some of the members of the National Commission and their consultants did not intend this result. Jonsen's *The Birth of Bioethics* contains some tracings of the deliberations of the National Commission and provides clues as to why the final report addresses the autonomy issue as it does.[9] While I am interested in understanding just why the final drafting of *Belmont* resulted in an inadequate autonomy principle, I also create the following conjectural history to suggest that the strong autonomy principle for which I advocate might be well-received by the authors of *Belmont* and those they consulted.

According to Jonsen, the National Commission worked through several drafts between the initial meeting at the Belmont House retreat and the final approval of the report on June 10, 1978. At the Belmont House retreat, the commissioners had in hand a small collection of background essays, most of them by ethicists. Drawing from these essays and their own deliberations, the commissioners devised seven principles, among which was the imperative to "respect self-determination."[10] These principles were then reformulated and pared down to three, among which was Engelhardt's "respect for persons as free moral agents." Although not a commissioner, Engelhardt had authored one of the background papers used by the commissioners, titled "Basic Ethical Principles in the Conduct of Biomedical and Behavioral Research Involving Human Subjects."[11]

One of Engelhardt's chief concerns was to indicate what is due to human subjects out of respect for them as free moral agents. He termed the idea of respect for persons as "a logical condition of morality," denoting something antecedent to the calculation of balancing harms and benefits involved in research. What Engelhardt sought was a strong, deontological principle that would be valid regardless of the benefits or burdens associated with specific research projects. In other words, he envisioned a principle that must be satisfied independent of the benefits from the research that might accrue either to the subjects or to society. Engelhardt's formulation of this rights-based principle was as follows: "One should respect human subjects as free moral agents out of a duty to such subjects to acknowledge their right to respect as free agents."[12] This principle was clearly too long to be fully incorporated in the report of a commission seeking a concise set of crisp, memorable phrases, and Engelhardt's formulation was abbreviated and carried forward in subsequent drafts of the National Commission as "respect for persons."[13] Left implicit in this abbreviated formulation is precisely what it is about persons that must be respected. My conjecture is that the truncation of "respect for human subjects as free moral agents" into "respect for persons" opened the door to the weak version of self-determination found in the final version of *Belmont*.

If this history has any merit, it means that the commissioners, or at least some of them, might well embrace the reformulation of *Belmont* that I suggest in the following section.

Revisions

My suggestions for revision of *Belmont* are twofold. First, "respect for persons" should be removed as the first principle and discussed in the introduction to the principles section of *Belmont* as the guiding vision for everything that follows. It is, in this sense, a preface to any and all ethical deliberations in the context of human subjects research. Second, "respect for autonomy" should become the first principle and its use as a principle explained and interpreted in the way suggested in the following discussion.

Respect for persons, as I have noted, should be thought of as a basic or foundational commitment guiding research with human subjects, rather than as one of three principles. The chief reason for this is that "respect" describes an attitude, or general demeanor. As such it tends to be vague and imprecise as a grounding for rules and actions.[14] The vagueness becomes evident when one asks exactly what it is about persons that is to be respected; the imprecision lies in not knowing just what would count as disrespect. For example, it is quite possible to say that one respects human subjects as persons and yet still not clearly honor their considered choices, which is, in effect, precisely the sort of action the illustration in *Belmont* would permit. By contrast, it would be a contradiction to say that one adheres to a principle of self-determination or autonomy for subjects and then ignore, override, or weigh on independent grounds the choices subjects make. To embrace respect for persons in the context of human subjects research is to affirm that subjects are human beings, of equal worth with the researchers, and that neither their illnesses nor their status as research subjects diminishes their humanity. Respect for persons is less an ethical principle than an ontological claim. Subjects are persons, after all, in a way parallel to the thesis Paul Ramsey forcefully incorporated into the title of his 1970 classic, *The Patient as Person.*[15]

This new situating of respect for persons within *Belmont* is supported by the widely varying interpretations of this term outside of *Belmont*. For some, this term is a shorthand version of Kant's second formulation of the categorical imperative, that is, "Act so that you treat humanity, whether in your own person or in that of another, always as an end and never as a means only."[16] Others associate respect for persons with Christian notions of self-effacing love and link it to duties owed to others.[17] This divergence of meaning stands in contrast to the principle of respect for autonomy, which has been more sharply defined. While the proper range of application of respect for autonomy is often contested, the meaning of this principle is not. Preserving respect for persons as an overarching value associated with a variety of moral traditions — and thus relieving it from the more definitive work of principles — will preserve *Belmont*'s broad appeal. Making respect for autonomy the first principle will provide the more precise focus needed to make sense of the exacting informed consent requirements found elsewhere in *Belmont* and in the Federal Common Rule.[18]

A second reason for such a revision follows from the importance of respect for persons as a basic insight undergirding the morality of the research enterprise generally. Respect for persons as a basic conviction is just as important to beneficence and justice as it is to autonomy. Beneficence has no proper goal unless we remember that it is persons we are seeking to protect and remove from harms. Justice is not a weighty principle unless we assume that all subjects in research are equally entitled to our respect as fellow human beings and that all should enter the calculus of distributing benefits and burdens fairly. No one is entitled to greater moral respect than others.

Some might suggest that moving respect for persons in the way I suggest would make it less prominent or important. In fact it does just the opposite. Placing respect for persons before the explication of the three principles would not only make it not subordinated to them but also establish that it is not confined only to the first principle. It is prolegomena to a proper understanding of all three principles. This fundamental moral assertion provides the necessary backdrop for interpreting the three principles in a potent way, a way that will undergird the rules and regulations rather than undermine them. Downie and Calman make a similar point, using a slightly different formulation, in their 1987 book *Healthy Respect: Ethics in Health Care*. They argue that the principle "respect for the autonomous person" is not like other principles in that the moral authority of the other principles "is derived from it, or it is presupposed by them."[19]

Respecting the autonomy of subjects has two cardinal components. First, it means treating subjects as free moral agents by actively promoting their independent reflection, deliberation, and decision making. This entails not only fulfilling the legal obligation of providing information necessary to an informed choice but also actively encouraging and enabling subjects to deliberate on the basis of their own values. Second, respecting autonomy in subjects means, quite simply, honoring the choices they make. No weighing, balancing, or adjudicating is required or appropriate.

For the sake of clarity and emphasis within *Belmont*, I would also suggest that the section concerned with protecting persons who lack decisional capacity, currently located under respect for persons, be relocated and placed under beneficence. This is where it belongs logically, because it concerns the obligation to prevent and minimize harms to the weak and vulnerable. Equally important, relocating this section would allow respect for autonomy to stand alone as a deontological norm. Moving the protective obligations into proximity with other duties of beneficence might reduce the temptation to see autonomy as a form of beneficence and increase the likelihood that self-determination will be understood as a distinctive, nonconsequentialist norm.

Consequentialist vs. Deontological Principles: Trumping and Balancing, in Context

One might object that I have misunderstood the nature and purpose of the *Belmont* principles, in terms of both their status and their function. For example, doesn't the existence of waivers of informed consent for some research involving medical records and for select categories of emergency research show that any autonomy principle is a consequentialist norm to be weighed and balanced, so that greater beneficence can justify some limitations on respect for autonomy? If autonomy were seen as a deontological norm, wouldn't it become an overriding or trumping principle, forcing all human subjects research into an ethical Procrustean bed, precluding all research in which informed consent is problematic?

This objection misses the point and confuses both the nature of deontological and consequentialist norms and how they must function together in a larger reasoning process. Although I believe, and will argue in the following discussion, that autonomy is a deontological norm, my chief criticism of the current formulation of *Belmont* is not that it portrays autonomy in a consequentialist way, but rather that it portrays autonomy in a weak (and possibly self-contradictory) way. Autonomy has little force if it places subjects in a passive role and makes their values and opinions subject to adjudication by some higher authority.

First, the existence of waivers of informed consent does not lend support to the idea that autonomy is a consequentialist principle; it indicates that it is not. A waiver would not be necessary if the value of consent were dependent on the good results it produces, for then its benefits could be weighed in a calculus, like other risks and benefits. When requirements to obtain consent are waived, they are not overbalanced by other goods, but are placed completely out of play, as in the examples given earlier, because efforts to obtain consent are not feasible. The duty to enter into informed consent conversations with subjects is not a duty grounded in the beneficial outcomes from so doing; the duty would still exist, when feasible, even if no good outcomes from this practice were in evidence. The waivers of informed consent may be justifiable because subjects cannot feasibly be asked, but not because the social benefits are so great or because we think subjects might not care or because too many subjects, if asked, would say no. The duty to obtain informed consent, grounded in respect for autonomy, is different in kind from the duty to seek and promote good outcomes from the research. Although benefits are sometimes thought to be so great as to override worries about the absence or imperfections of consent, anticipated or actual benefits from the research cannot *substitute* for such absence or imperfections. Consent serves distinctively nonconsequentialist purposes, the most basic of which is to enhance the autonomy and thereby protect the dignity of the subject. Of course, protecting the dignity and enhancing the autonomy of subjects may produce a good outcome, such an increased moral self-understanding or decisions that subjects can more easily live with when things go wrong in the research endeavor. But these consequentialist goods are surplus, or bonus, goods and do not alter the deontological nature of consent as a vehicle for goods that are intrinsically valuable.

Second, the claim that autonomy is a deontological principle is not a claim for its dominant status or trumping ability. A deontological principle is not necessarily more powerful than a consequentialist one. Rather it denotes values that cannot be measured in outcomes. To be sure, we might think of some contexts in which respect for autonomy is so important that it works like a trumping principle. For example, ordinarily in Phase 1 oncology research, informed consent is a necessary but not sufficient condition for enrolling subjects; without consent the research should not be conducted. But not all research protocols involve subjects who possess optimal capacity for engaging in informed consent. In these situations, researchers and Institutional Review Boards (IRBs) will have to judge the relative importance of a full and optimal consent process along with other factors in deciding whether research is permissible. For example, in protocols involving persons of marginal or compromised reasoning capacity — such as children or some persons with mental illnesses — it may be decided that the benefits of some research projects are of such magnitude that a less-than-optimal consent is acceptable, so long as full protections and safeguards are in place. Here, the absence of a full and complete informed consent process does not trump the deliberations but becomes one ingredient among several to consider. The likely quality of the consent process is weighed against other factors relevant to the overall ethics of the research, for example, minimal risk and significant societal benefit or a high potential for direct benefit to subjects. Autonomy as a deontological norm is balanced against not only the utilitarian calculus of social benefits but also other deontological norms, such as the duty to protect subjects. Final determinations about the ethical permissibility of any given research project lie less in the nature of the principles — as deontological or consequentialist — and far more in careful judgments made by considering all the factors involved.

Note that these examples are, again, ones in which autonomy is less feasible and in which outcomes do not substitute for consent but simply outweigh it. (This is, of course, *not* to say that in every case in which consent is problematic, potential benefits will be a decisive factor.) Note also the way autonomy functions in these difficult cases. Autonomy is not *measured by* beneficial outcomes, but its importance is *measured against* the potential for such outcomes, among other things. To make autonomy a trumping principle in all situations would make research with some populations impossible, and that is not my aim, nor should it be the aim of research ethics more generally. My aim here is to ensure that when an imperfectly realized autonomy (measured in the quality of consent) must be weighed against other principles, both deontological and consequentialist, a sufficiently vigorous form of the autonomy principle is used.

Regression to the Past?

This chapter may evoke the objection that the *Belmont Report* has served very well for more than twenty-five years. Perhaps it could be improved here or there, but it isn't seriously broken. Why modify it for what some might regard as cosmetic alterations? One might argue that the explication of informed consent in other parts of *Belmont* bespeak a strong autonomy principle, so that there is little doubt about what the commissioners intended. Moreover, great strides have been made in the field of research ethics, and most investigators seem to understand their obligation to obtain informed consent and to honor the autonomy of subjects, in spite of *Belmont's* flawed interpretation of the principle of autonomy.

First, it should be pointed out that *Belmont* serves as the benchmark against which IRBs and individual investigators assess the moral probity of their work. It serves as both a guide for current research activities and the standard against which to measure any new regulations. *Belmont* also serves as a basic ethical template for the education of future researchers. The prominence and authority of *Belmont* impose an obligation for periodic, critical reexamination to be certain that it reflects the best ethical thinking we can muster.

Second, although there is an extensive explication of informed consent as an embodiment of the autonomy principle in *Belmont*, the obligation to promote and enable an active role for subjects is not emphasized. Rather, in the applications section of *Belmont*, what is emphasized is giving subjects "the opportunity to choose." In a similar way, researchers are given responsibility for "ascertaining that the subject has understood the information."[20] This is fine as far as it goes, but assumptions of subject passivity still underlie these phrases, and no effort to move subjects from this posture is enjoined.

Third, it is far from clear that research practices are uniformly guided by a strong autonomy principle. Recent reports from the General Accounting Office,[21] the Recombinant DNA Advisory Committee,[22] and the Advisory Committee on Human Radiation Experiments[23] indicate that there is confusion among both patients and clinicians about what separates clinical research from therapeutic practices. Accompanying this confusion is a tendency by both clinical investigators and patients to see clinical research as another form of therapy and to apply inappropriately the ethical norms that shape physician-patient interaction to investigator-subject relationships.[24]

This tendency to blur the differences between research and therapy and to collapse research ethics into the norms of medical practice is remarkable because of its similarities to the understanding of research ethics dominant in the pre-*Belmont* era. For example, Rothman has described American research ethics in the twenty-year period immediately following World War II as one in which "the ethos of the examining room cloaked the activities of the laboratory, and the trust accorded the physician encompassed the researcher."[25] Faden, Beauchamp, and King, in *A History and Theory of Informed Consent*, also claim that between 1946 and 1966 the norms of beneficence that applied to physicians were simply transferred to researchers and that, without a counterbalancing principle of autonomy, the chief concern in human subjects research was to control risks rather than to ensure the rights of subjects to autonomous choice. In many cases, the exclusive means of protecting subjects from risk was simply reliance on the moral character of investigators.[26] In the absence of a clear and robust principle of respect for autonomy, contemporary research ethics risks regression to these inadequate norms of the past. A recent article by Truog, Robinson, Randolph, and Morris in the *New England Journal of Medicine*[27] suggests that the tendency to blur research and therapy in the clinical setting is still a frequent occurrence, with serious consequences for autonomy and consent requirements.

The current environment for research is marked by social and medical pressure to open up clinical trials to more individuals seeking access to treatment, especially to patients for whom existing remedies have been ineffective. Patient advocacy groups have increasingly demanded access to trials when such studies seemed to provide their best hope. In a complementary way, the sympathy of clinicians for the seriously ill has resulted in "easing the tension between the roles of 'patient' and 'subject'"[28] and a corresponding loosening of federal research regulations designed to help desperate patients. This accords with a long-standing medical tradition of seeking unproven remedies for patients who seem to have no other option. This climate is in marked contrast to the attitudes about research prominent in the 1970s when, in the wake of revelations about the U.S. Public Health Service's trampling of human rights in the Tuskegee syphilis experiments, the interest in protecting human subjects and ensuring their autonomy was paramount. Protecting human subjects and ensuring their autonomy remains a high priority, but these goals now compete with the aim of opening clinical trials to desperate patients under the warrant of therapeutic beneficence.

Given the increasing climate of enthusiasm about medical research, it is important to affirm a strong, clear, and distinct principle of respect of autonomy for human subjects. Recent funding increments for the National Institutes of Health, with a commensurate increase in human subjects research, underscore the need for a robust autonomy principle to undergird federal regulations. Such a principle will help to ensure that beneficent intentions do not overshadow self-determination and that scientific ambitions and therapeutic zeal will function within boundaries of voluntary and well-informed subject participation. The absence of such boundaries would be a loss for the freedom and dignity of human subjects, but also a loss for the ethical sensibilities of clinical investigators and eventually for the public's confidence in scientific inquiry in medicine. If our aim is to demonstrate respect for persons, respecting the autonomy of subjects is the essential first step.

Notes

I thank Myra L. Collins, Arlene M. Davis, Stuart G. Finder, Nancy M. P. King, Stephen Pemberton, Daniel G. Nelson, Robert J. Levine, Christy Parham-Vetter, Giles Scofield, and Keith A. Wailoo for their comments and suggestions on earlier versions of this manuscript. Support for this work was provided by the National Human Genome Research Institute (Grant No. 1-RO1-HGo1177-01A2).

1. Office for Protection from Research Risks, U.S. Department of Health, Education and Welfare, *The Belmont Report* (Washington, DC: OPRR Reports, ACHRE No. HHS-011795-A-2, April 18, 1979), 4 [hereafter *Belmont Report*].

2. Ibid.

3. Ibid.

4. Tom L. Beauchamp and James F. Childress, *Principles of Biomedical Ethics*, 4th ed. (New York: Oxford University Press, 1994), 125.

5. Ibid., 127.

6. *Belmont Report*, 4.

7. See, e.g., Larry R. Churchill, "Getting from 'I' to 'We,'" in *A Good Old Age? The Paradox of Setting Limits*, ed. Paul Homer and Martha Holstein, 109–19 (New York: Simon and Schuster, 1990).

8. See, among many other studies and reports, Jonathan Moreno et al., "Updating Protections for Human Subjects Involved in Research," *Journal of the American Medical Association* 280 (1999): 1951–54.

9. Albert R. Jonsen, *The Birth of Bioethics* (New York: Oxford University Press, 1998), 102–4. It is interesting that at the National Bioethics Advisory Commission's "*Belmont* Revisited" conference in Charlottesville, Virginia, on April 16–18, 1999, at which this paper was presented, there was substantial disagreement among the authors and contributors about the process of writing and editing the various drafts of the *Belmont Report*, including just who contributed portions of the final document.

10. Jonsen, *Birth of Bioethics*, 103.

11. *Belmont Report*, appendix, vol. 1, DHEW Pub. No. (OS) 78-0013, 8-1– 8-45.

12. *Belmont Report*, appendix, vol. 1, 8-5.

13. Jonsen, *Birth of Bioethics*, 103.

14. John E. Atwell has made a similar point in his essay "Kant's Notion of Respect for Persons," in *Respect for Persons*, Tulane Studies in Philosophy, vol. 31 (New Orleans, LA: Tulane University Press, 1982). Atwell's essay opens with a critique of Downie and Telfer's sweeping claims in *Respect for Persons* (London: George Allen and Unwin, 1969), for the foundational status of "respect for persons" as a principle. Atwell argues that "practical moralists have made appeal to respect for persons for the sake of supporting almost every imaginable policy — capital punishment *and* its abolition, abortion on demand *and* the 'right to life,' etc." Atwell notes that Downie and Telfer claim to derive their principle from Kant, but he argues that Kant's concerns are better expressed as something like "reverence for humanity" (17ff.).

15. Paul Ramsey, *The Patient as Person: Explorations in Medical Ethics* (New Haven, CT: Yale University Press, 1970).

16. Immanuel Kant, *Foundations of the Metaphysics of Morals*, trans. by Lewis White Beck (New York: Macmillan, 1985), 47.

17. See the entry "Respect for Persons," authored by Gene Outka in *The Westminster Dictionary of Christian Ethics*, ed. James F. Childress and John Macquarrie (Philadelphia, PA: Westminster Press, 1986), 541–45.

18. Office for Protection from Research Risks, U.S. Department of Health and Human Services, *Protection of Human Subjects*, Title 45, Code of Federal Regulations, Part 46, revised June 18, 1991, at 46.116.

19. R. S. Downie and K. C. Calman, *Healthy Respect: Ethics in Health Care* (London: Faber and Faber, 1987), 50. See also Downie and Telfer, *Respect for Persons*, who argue that respect for persons is "the paramount moral attitude" and that "all other principles are to be explained in terms of it" (p. 33). This is perhaps overstating the case (see note 14). I would agree that respect for persons is a central commitment that should guide research with human subjects, but to suggest that all other principles should be explained in terms of it begs too many issues. I would say rather that the interpretation of all the other principles of *Bel-*

mont are enhanced when we make respect for persons a central attitude or conviction informing all aspects of human subjects research.

20. *Belmont Report,* 5–6.

21. General Accounting Office, *Scientific Research: Continued Vigilance Critical to Protecting Human Subjects* (Washington, DC: General Accounting Office, GAO/HEHS-96-72, March 8, 1996). See especially p. 23. The report contends that "the line between research and treatment is not always clear to clinicians" and that it is controversial whether certain interventions should be categorized as research.

22. Gail Ross et al., "Gene Therapy in the United States: A Five-Year Status Report," *Human Gene Therapy* 7 (1996): 1789. The closing statement reads: "In sum, while the public has anticipated that this new form of therapy will lead to novel medical cures, it is still too early to tell if and when gene therapy will achieve its goals."

23. Advisory Committee on Human Radiation Experiments, "Subject Interview Study," in *The Human Radiation Experiments: Final Report of the President's Advisory Committee* (New York: Oxford University Press, 1996), 459–81. See especially pp. 468–70. Notably, the Subject Interview Study found that patients did not readily distinguish between research and medical treatment. Sixty-seven percent of patients said that the hope of getting better treatment "contributed a lot" to their decision to join a research project. Many others said that because of the seriousness of their illnesses they felt they had "little choice" but to enroll in a research project.

24. For a detailed examination of how confusion between research and therapy has affected attitudes and policies about "gene therapy" research, together with recommendations for remedying these problems, see Larry R. Churchill, Myra L. Collins, Nancy M. P. King, Stephen G. Pemberton, and Keith A. Wailoo, "Genetic Research as Therapy: Implications of 'Gene Therapy' for Informed Consent," *Journal of Law, Medicine and Ethics* 26 (1998): 38–47.

25. David J. Rothman, *Strangers at the Bedside: A History of How Law and Bioethics Transformed Medical Decision Making* (New York: Basic Books, 1991), 66.

26. Ruth R. Faden and Tom L. Beauchamp, with Nancy M. P. King, *A History and Theory of Informed Consent* (New York: Oxford University Press, 1986), chapters 5 and 6.

27. Robert D. Truog, Walter Robinson, Adrienne Randolph, and Alan Morris, "Is Informed Consent Always Necessary?" *New England Journal of Medicine* 340 (1999): 804–6.

28. Steven Epstein, *Impure Science: AIDS, Activism, and the Politics of Knowledge* (Berkeley: University of California Press, 1996), 215–16.

9 The National Commission's Ethical Principles, With Special Attention to Beneficence

Robert J. Levine

In this chapter I shall offer some comments on the ethical principles identified as basic or fundamental by the National Commission for the Protection of Human Subjects of Biomedical and Behavioral Research (hereafter the National Commission). Then I shall comment more specifically on the fundamental principle of beneficence.[1]

The act that established the National Commission charged it to "conduct a comprehensive investigation and study to *identify* the basic ethical principles which should underlie the conduct of biomedical and behavioral research involving human subjects."[2] [emphasis added]. It is important to note that the National Commission was charged to identify, not to invent, the basic ethical principles. To do this they looked back at various national and international codes and regulations with the aim of detecting ethical principles expressed explicitly or implicitly in these documents. They also commissioned various consultants to assist in this project.

The National Commission defined a "basic ethical principle" as a "general judgment that serves as a basic justification for the many particular prescriptions for and evaluations of human actions."[3] A "fundamental ethical principle" is one that, within a system of ethics, is taken as an ultimate foundation for any second-order principles, rules, and norms; a fundamental principle is not derived from any other statement of ethical values.[4] The National Commission identified three fundamental ethical principles as particularly relevant to the ethics of research involving human subjects: respect for persons, beneficence, and justice. The norms and procedures presented in regulations and ethical codes are derived from and are intended to embody and uphold these fundamental principles. In an early draft of the *Belmont Report*, the National Commission expressed its view of the general applicability of these ethical principles:

> Reliance on these three fundamental underlying principles is consonant with the major traditions of Western ethical, political and theological

thought presented in the pluralistic society of the United States, as well as being compatible with the results of an experimentally based scientific analysis of human behavior.[5]

Thus, in view of the National Commission's findings, these principles pertain to human behavior in general; it is through the development of norms that they are made peculiarly relevant to specific classes of activities such as research and the practice of medicine.

The National Commission did not embrace any particular foundational ethical theory such as deontology or consequentialism, even though the names of their principles suggest that they did. In particular, "respect for persons" suggests deontological roots, and "beneficence" suggests utilitarianism. Some commentators have written that the National Commission failed to specify a foundational theory for their deliberations and conclusions. I believe that the proper interpretation in this regard is that the National Commission considered this possibility and rejected it as unnecessary. Implicit in the publications of the National Commission is an appreciation of the validity of both deontological and consequential reasoning. Albert Jonsen and Stephen Toulmin, who participated in the deliberations of the National Commission, have argued that much of the commission's work resembled casuistry.[6]

A hierarchical-deductive model of ethical reasoning was fashionable in the 1970s. As far as I can determine, the National Commission did not state whether it employed a model that was hierarchical, deductive, or inductive. Several commissioners and members of the staff who wrote of the National Commission's ethical reasoning have assumed a hierarchical-deductive model.[7] In such a model, norms — for example, the requirement for informed consent — are deduced from abstract principles — in this case, the principle of respect for persons.

The three ethical principles identified as fundamental by the National Commission were envisioned by that commission as having, at least in the abstract, equal moral force. There was no intent on the part of the National Commission to establish a lexical ordering.

The writings of many commentators in the years since the National Commission published the *Belmont Report* are grounded in assumptions that are inconsistent with the commission's vision. Most prominently:

- Authors assume that there is or ought to be a lexical ordering of the principles. Those who mistakenly attribute this view to the National Commission most commonly assert that the principle of respect for persons may be thought of as trumping beneficence. Some commentators compound this error by offering an incorrect reason for this alleged trumping: it is, so they say, because principles that are derived deontologically take priority over those that are based on consequential reasoning.[8]
- Fundamental principles are treated as if they are direct action guides. Commentators state that in a particular situation or context, benefi-

cence requires a particular action rather than simply referring to the relevant norm — the true action guide. This error tends to oversimplify what should be recognized as highly complicated moral judgments — for example, when norms related to (e.g.) beneficence give rise to inconsistent or incompatible judgments about particular actions that ought to be performed in particular cases.

- There is a widespread tendency to equate respect for persons, as this principle was defined by the National Commission, with principles called autonomy or respect for autonomy. This has a propensity to obscure the fact that there is much more to persons than a capacity to be self-determining. There are attributes of persons other than autonomy that create a moral obligation to treat them with respect; moreover, treating persons with respect requires a responsiveness to attributes other than autonomy.[9]

Beneficence

In the *Belmont Report*, the National Commission had this to say about the principle of beneficence:

> The term, beneficence, is often understood to cover acts of kindness or charity that go beyond strict obligation. In this document, beneficence is understood in a stronger sense, as an obligation. Two general rules have been formulated as complementary expressions of beneficent actions in this sense: 1) Do no harm and 2) maximize possible benefits and minimize possible harms.[10]

The principle of beneficence is firmly embedded in the ethical tradition of medicine. It is often said that the first principle of medical ethics is *primum non nocere*, a literal translation of which yields "first do no harm" or "above all, do no harm." It is further said that this principle was first articulated by Hippocrates as part of the Hippocratic oath, but this is incorrect in two respects. The closest approximation of this statement in the entire Hippocratic corpus is found, not in the oath, but in a book titled *Epidemics*: "As to diseases, make a habit of two things — to help, or at least to do no harm."[11] To the extent that moderns believe that medical ethics is or ought to be influenced by the writings attributed to Hippocrates, we may consider ourselves fortunate. If the first principle of medical ethics were truly "above all, do no harm," this would rule out almost all modern therapeutics. Administration of almost any therapeutic regimen is associated with some risk of harm. The accurate quotation of the Hippocratic writings is much more compatible with modern notions of risk-benefit balancing.

Some commentators believe that separation of these two obligations into two discrete ethical principles, beneficence (do good) and nonmaleficence (do no harm), would tend to decrease confusion.[12] Others, in harmony with the National Commission, envision these two ethical obligations as components of a single principle named beneficence. One such commentator is William Frankena, who identifies four distinct obligations that derive from the principle of beneficence.[13] Listed in decreasing order of moral force, these obligations follow:

A. One ought not to inflict evil or harm.
B. One ought to prevent evil or harm.
C. One ought to remove evil.
D. One ought to do or promote good.

Statement A is a straightforward articulation of the principle of non-maleficence. Few, if any, would argue that the injunction against inflicting harm or evil is not a very strong prima facie duty. Statement D, by contrast, is not regarded generally as a duty or obligation but rather, as the National Commission pointed out, an exhortation to act kindly or charitably. Statement D becomes a duty in the strict sense when one consents or contracts to be bound by it. Physicians, for example, pledge themselves to act for the benefit of patients when they take the Hippocratic oath or one of its modern revisions.

The National Commission identified beneficence as a strict obligation for several reasons:

- Research entails exposing persons to risk of injury without compensating benefit to the individual research subject. Thus, there must be some other justification for the imposition of such risk. The most acceptable justification is the pursuit of benefits for the collective. When one or more components of the research protocol hold out the prospect of direct benefit for the individual research subject, the necessary justification of the risk in terms of anticipated benefit is more complicated.
- Even when there is no risk, it is improper to waste the subjects' time in pointless pursuits.
- Ordinarily, during the process of informed consent, the investigator makes a commitment to the prospective subject to pursue the acquisition of new knowledge that will be beneficial to the collective.
- There is a duty to the financial sponsors of the research — often an agency of the federal government — to spend their money well. There is in almost every case a written agreement with the financial sponsor in the form of a grant or contract that specifies the investigator's commitment to pursue benefits for the collective.

The principle of beneficence, as interpreted by the National Commission, creates an obligation to secure the well-being of the individuals who serve as research subjects and to develop information that will form the basis of being better able to serve the well-being of similar persons in the future. However, in the interests of securing societal benefits, one should not intentionally injure any individual. Calling upon persons to agree to accept risks of injury in the interest of producing benefits for others is morally less offensive than taking an action that will certainly injure an identifiable individual. As I shall discuss in detail, research tends to present us with much more complex risk-benefit calculations than do most classes of medical practice.

What does it mean to secure the well-being of research subjects? First, it means refraining from any actions that will certainly injure a research subject; this is what the National Commission means when it specifies that one of the rules embodied in the principle of beneficence is "do no harm." It does not mean "above all, do no harm." In addition to that, the leading codes of research ethics specify that investigators must be competent to provide "care" for the research subject. The Nuremberg Code states, "The highest degree of skill and care should be required through all stages of the experiment of those who conduct or engage in the experiment." To this Helsinki adds, "The responsibility for the research subject must rest with a medically qualified person."

Competence to care for the subjects of much biomedical research requires that at least one member of the research team be responsible for observing the subject with a view toward early detection of adverse effects of his or her participation or other evidence that the subject should be excluded from the study. The investigator should have the competence to assess the subjects' symptoms, signs, and laboratory results, as well as to intervene as necessary in the interests of minimizing any harm, for example, prompt administration of an antidote to a toxic substance.

The obligation to develop information that will form the basis of our being better able to secure the well-being of persons in the future may be considered a defining attribute of biomedical research. Research is defined in the *Belmont Report* as a class of activities designed to develop or contribute to the development of generalizable knowledge. Generalizable means that the knowledge to be developed will be applicable to people in general, not simply to serve the health interests of particular people as in the practice of medicine or of other health care professions.

Justification of Risk

The National Commission concluded that risk to research subjects should be minimized by using the least risky procedures that are consistent with sound research design and, when feasible, by using procedures already being performed on the subjects for clinical indications. The latter class of activities, often called opportunistic research, is exemplified by using as research material leftover body fluids (e.g., blood or spinal fluid) obtained for diagnostic purposes. It is also exemplified by obtaining an additional few milliliters of spinal fluid or coronary sinus blood for research purposes from patients who require a spinal tap or coronary sinus catheterization for diagnostic purposes rather than performing such invasive procedures for research purposes on individuals who otherwise would not require one.

The National Commission further concluded that risks must be justified by being in a reasonable or favorable relationship to the anticipated benefits. This conclusion represented a departure from the standard that had been established earlier in federal regulations. Regulations promulgated in 1974 by the Department of Health, Education and Welfare (DHEW) required the Institutional Review Board (IRB) to determine that

> the risks to the subjects are so outweighed by the sum of the benefit to the subject and the importance of the knowledge to be gained as to warrant a decision to allow the subject to accept these risks.[14]

The National Commission's recommendation on this point was decidedly less paternalistic:

> If the prospective subjects are normal adults, the primary responsibility of the IRB should be to assure that sufficient information will be disclosed in the informed consent process, provided the research does not present an extreme case of unreasonable risks.

How does one find that the balance of hoped-for benefits is in a favorable relation to the risks to the individual subject? It is necessary to acknowledge two inherent limitations in our ability to answer this question precisely. First, it is not possible to construct a satisfactory algorithm for determining what is favorable or reasonable. Favorable and reasonable are dispositional attributes, and attempts to pin down the meanings of such attributes in rigidly defined terms will, if successful, only defeat their purpose. Favorable, then, is a term used to suggest to reasonable persons that there is something about the balance of harms and benefits that other reasonable persons are likely to find favorable. Second, judgments about the balance of harms and benefits are expected of at least three classes of agents or agencies: these are (1) investigators, in order to justify their proposing to do the research; (2) IRB members, in order to decide whether to approve it; and (3) subjects, in order to decide whether to consent to or to refuse participation in research. It is unsurprising that these three classes may reach differing judgments about the favorableness of the balance of harms and benefits presented by any particular research proposal.

Justification of risk is never an isolated event. To say that the imposition of risk in research is justified presupposes that the plan to do the research is also in accord with requirements of all relevant ethical norms and procedures.

In research, all risks (of injury) are borne by the research subjects, while at least some and at times all of the benefits will be for the collective. The determination that risks and benefits are in a reasonable relation to one another is often complicated. Risks associated with the pursuit of the individual subject's well-being are to be justified according to different standards than are risks associated with the pursuit of benefit to the collective.

Before the National Commission's deliberations on this topic, it was customary to classify research into two categories: therapeutic and nontherapeutic research. Each of these categories, then, had its own criteria for justification. Standards for justification of therapeutic research were less stringent than those for nontherapeutic research. In general, any research having one or more therapeutic components was classified as therapeutic and justified accordingly. This led to what I have labeled the fallacy of the package deal, because the many nontherapeutic components of such research protocols were also justified according to the relatively relaxed standards designed for therapeutic research.[15]

The National Commission explicitly repudiated the distinction between therapeutic and nontherapeutic research. It replaced this distinction with a much more rational system in which the various components of the research protocol are classified in terms of whether or not they are "interventions or procedures that hold out the prospect of direct benefit for the individual subject."[16] Those that do may be called beneficial or therapeutic modalities; their justification closely approximates the justification of therapeutic interventions in the practice of medicine. It is a strictly personal felicific calculation. The only ceiling for the probability or magnitude of risk from beneficial procedures is that they are not to exceed those of the benefits that can be reasonably expected. One additional criterion is that the relationship of anticipated benefit to the risk presented by the modality must be at least as advantageous to the subject as that presented by any available alternative, unless, of course, the individual has considered and refused to accept a superior alternative.

Nonbeneficial procedures or interventions are justified much differently. For example, no more than minimal risk may be presented to a child without special justification. If it is more than a minor increment above minimal risk, review at a national level is required. If it is only a minor increment, then the procedure must be justified in terms such as the following: the procedure itself must present an experience to the child-subject that is reasonably commensurate with those inherent in his or her actual or anticipated experience. Moreover, there is a more stringent standard for evaluating the anticipated contribution to generalizable new knowledge.

In his book *The Birth of Bioethics*, Albert Jonsen identifies "four major contributions to understanding the ethics of research involving human subjects" made by the National Commission. Among these four he gives pride of place to its definition of research, which "abolishes the long-cherished but misleading distinction between therapeutic and nontherapeutic research."[17]

Since the 1970s, the distinction between therapeutic and nontherapeutic research has been systematically purged from federal regulations. The last step in this process has been to remove it from subpart B, the regulations for research involving fetuses, pregnant women, and in vitro fertilization. These regulations were based on the one report the National Commission published before it accomplished its conceptual clarifications.[18] In addition, for a while we appeared to be on the verge of seeing this illogical concept removed from the Declaration of Helsinki,[19] the document that seemed to establish it in the lexicon of research ethics in the first place, but this has not yet occurred.[20] If and when it does occur, we will have witnessed the removal of an obstacle to the proper understanding and application of the principle of beneficence.

Notes

This work, which is based on a presentation to the symposium "Belmont Revisited," at the University of Virginia, April 17–18, 1999, was funded in part by grant number PO1 MH/DA 56 826-01A1 from the National Institute of Mental Health and the National Institute on Drug Abuse, grant number 1 P30 MH 62294 01A1 from the National Institute of Mental Health, and a grant from The Patrick and Catherine Weldon Donaghue Medical Research Foundation.

1. For further elaboration and documentation of the ideas presented in this chapter, see Robert J. Levine, *Ethics and Regulation of Clinical Research*, 2nd ed. (New Haven, CT: Yale University Press, 1988). Some passages of this chapter are adapted or abridged from that book.

2. The National Commission was established in 1974 by Title II of the National Research Act (Public Law 93-348).

3. The National Commission for the Protection of Human Subjects of Biomedical and Behavioral Research, *The Belmont Report: Ethical Principles and Guidelines for the Protection of Human Subjects Research*. DHEW Publication No. (OS) 78-0012, 1978, 4.

4. The National Commission generally used the term "basic" and sometimes "fundamental"; while I prefer the latter, I use *basic* when quoting a passage in which the National Commission used that term.

5. Robert J. Levine, "On the Relevance of Ethical Principles and Guidelines Developed for Research to Health Services Conducted or Supported by the Secretary," in The National Commission for the Protection of Human Subjects of Biomedical and Behavioral Research, *Report and Recommendations: Ethical Guidelines for the Delivery of Health Services by DHEW*, DHEW Publication No. (OS) 78-0011, 1978, appendix, pp. 2.1–2.36.

6. Albert R. Jonsen and Stephen Toulmin, *The Abuse of Casuistry: A History of Moral Reasoning* (Berkeley: University of California Press, 1988). In my opinion, if the National Commission employed casuistry, it was not aware of its having done so at the time. However, I agree with Jonsen and Toulmin that there is a widespread tendency among groups (and individuals) who are responsible for making decisions in particular cases to think carefully about how or what they decided the last time they were presented with a similar case. There is also a tendency to try to make present decisions appear to be consistent with past decisions unless one can demonstrate that the present case is different in important and relevant respects from the case or cases that established the precedent or paradigm. In these respects the deliberations of the National Commission resembled casuistry.

7. See, e.g., Tom L. Beauchamp and James F. Childress, *Principles of Biomed-*

ical Ethics, 5th ed. (New York: Oxford University Press, 2001), as well as earlier editions; Karen A. Lebacqz and Robert J. Levine, "Respect for Persons and Informed Consent to Participate in Research," *Clinical Research* 25 (1977): 101–7.

8. By saying that they compound the error, I mean that these commentators assume incorrectly that the National Commission considered its principles either deontologically or consequentially derived. At this time I do not intend to argue or assert that deontological principles are either superior or inferior to consequential principles.

9. One attribute of persons that both creates an obligation to treat them with respect and defines ways in which they are to be treated with respect is relationality, their capacity for developing relationships with other human beings. See, e.g., Carol Gilligan, *In a Different Voice: Psychological Theory and Women's Development* (Cambridge, MA: Harvard University Press, 1982); Karen A. Lebacqz, *Professional Ethics: Power and Paradox* (Nashville, TN: Abingdon Press, 1985). Although an obligation to show respect for relationality was not clearly explicated in the *Belmont Report*, it is implicit in the National Commission's report on research involving children. Respect for relationality is one of the basic premises that ground the moral obligation to maintain the confidentiality of private information. See, e.g., Sissela Bok, *Secrets: On the Ethics of Concealment and Revelation* (New York: Pantheon, 1982).

10. *The Belmont Report*, 7–8.

11. See Albert R. Jonsen, "Do No Harm," *Annals of Internal Medicine* 88 (1978): 827–32.

12. See, e.g., Beauchamp and Childress, *Principles of Biomedical Ethics*.

13. William K. Frankena, *Ethics*, 2nd ed. (Englewood Cliffs, NJ: Prentice Hall, 1973).

14. Section 46.102.

15. For further discussion of the problems presented by the use of the illogical distinction between therapeutic and nontherapeutic research, see Robert J. Levine, "The Need to Revise the Declaration of Helsinki," *New England Journal of Medicine* 341 (1999): 531–34, and *Ethics and Regulation of Clinical Research*, 8–10.

16. 45 C.F.R. 46 Subpart B.

17. Albert R. Jonsen, *The Birth of Bioethics* (New York: Oxford University Press, 1998).

18. 45 C.F.R. 46 Subpart B — "Additional Protections Pertaining to Research, Development and Related Activities Involving Fetuses, Pregnant Women and Human In Vitro Fertilization" was based on the National Commission's *Research on the Fetus: Report and Recommendations*, DHEW Publication No. (OS) 76-127, 1975. It is disappointing that the National Bioethics Advisory Commission reintroduced the concept in its report on research involving persons with mental disorders that may affect decision-making capacity. See Robert J. Levine, "Commentary on the Decision by the National Bioethics Advisory Commission to Employ the Illogical Distinction between Therapeutic and Nontherapeutic Research," *BioLaw* 2 (1999): S424–48.

19. See Levine, "The Need to Revise the Declaration of Helsinki."

20. My optimism with regard to purging the distinction between therapeutic and nontherapeutic research from the Declaration of Helsinki seemed well-founded when the paper from which this chapter is derived was first presented. What actually happened, however, was a disappointment. The World Medical Association, in its most recent revision of the Helsinki Declaration (52nd WMA General Assembly, Edinburgh Scotland, October 2000), removed the language of therapeutic and nontherapeutic. However, in some of its articles it retained the concept as an organizing principle. For example, there is now a Section C subtitled "Additional Principles for Medical Research Combined with Medical Care." In it there is Article 28, which reads:

> The physician may combine medical research with medical care, only to the extent that the research is justified by its potential prophylactic, diagnostic or therapeutic value. When medical research is combined with medical care, additional standards apply to protect the patients who are research subjects.

This article continues to rule out research in the fields of pathogenesis, pathophysiology, and epidemiology, among other things. For further discussion of this topic, see Robert J. Levine, "Placebo Controls in Clinical Trials of New Therapies for Conditions for Which There Are Known Effective Treatments," in *The Science of the Placebo: Toward an Interdisciplinary Research Agenda*, ed. H. A. Guess, A. Kleinman, J. W. Kusek, and L. W. Engel, 264–80 (London: BMJ Books, 2002).

10 Justice beyond *Belmont*

Patricia A. King

In the years immediately preceding the establishment of the National Commission for the Protection of Subjects in Biomedical and Behavioral Research in 1974, revelations of unethical research, and congressional and public reaction to these revelations, resulted in the initiation of federal regulatory oversight to protect the rights and welfare of research subjects. The National Commission's effort continued and advanced the process of protecting research subjects. Although the National Commission's work emphasized respect for persons and autonomous decision making as means of protecting subjects, it also recognized the importance of the principles of beneficence and justice in promoting this goal. As one commentator stated, "Our basic approach to the ethical conduct of research . . . was born in scandal and reared in protectionism."[1]

This chapter focuses on the principle of justice and its corresponding application, selection of subjects. First, I describe how the National Commission elaborated on the principle of justice and selection of subjects and conclude that this formulation of the principle of justice was influenced by its desire to protect vulnerable research subjects from coercion and exploitation and to avoid a mismatch between those who were the subjects of research and those who were its primary beneficiaries. Next, I examine the experiences of the last thirty years that have served, for the most part, to redirect concern away from the potential for unethical and risky research to the possibilities for individual and societal health and well-being that research brings. This shift, which emphasizes therapeutic research rather than nontherapeutic research, access rather than protection, and benefits rather than risks and burdens, is significant to our theoretical and practical understanding of the principle of justice. I close by identifying some of the implications of this dramatic change. An important feature of this shift is that it highlights the need to consider the relevance of group membership in fair distribution of the benefits and burdens of research. Groups vulnerable thirty years ago remain vulnerable today. The challenge is to devise research policies and practices that are both inclusive

136

and protective. I further suggest that compensatory and procedural justice, in addition to distributive justice, are relevant if this challenge is to be met.

Albert Jonsen, a member of the National Commission, has written that "justice is the neglected sibling among the principles of bioethics; always acknowledged but seldom given significant tasks or much praise."[2] I am optimistic that in the future the principle of justice will play a more critical role than it has in the past.

The National Commission's Formulation of the Principle of Justice

The National Commission's work focused on two aspects of justice that it considered relevant to research with human subjects:

> Who ought to receive the benefits of research and bear its burdens? This is a question of justice, in the sense of "fairness in distribution" or "what is deserved." An injustice occurs when some benefit to which a person is entitled is denied without good reason or when some burden is imposed unduly. Another way of conceiving the principle of justice is that equals ought to be treated equally.[3]

Thus, the National Commission viewed the principle of justice in the research context as relevant to selection of subjects. It ignored other aspects of justice, notably compensatory justice, that arguably were relevant to its task of protecting research subjects. Compensatory justice might have been relevant to addressing claims of injured research subjects for redress from harm or injury.

The *Belmont Report* took note of historical events to show the connection between justice and research involving human subjects. It found that, historically, members of some groups tended to bear the burdens of participation in research:

> During the 19th and early 20th centuries the burdens of serving as research subjects fell largely upon poor ward patients, while the benefits of improved medical care flowed primarily to private patients. Subsequently, the exploitation of unwilling prisoners as research subjects in Nazi concentration camps was condemned as a particularly flagrant injustice. In this country, in the 1940s, the Tuskegee syphilis study used disadvantaged, rural black men to study the untreated course of a disease that is by no means confined to that population.[4]

Moreover, the National Commission explicitly recognized that the focus of ethical research policies had to operate at both the individual and group levels.

> Individual justice in the selection of subjects would require that researchers exhibit fairness: thus, they should not offer potentially benefi-

cial research to some patients who are in their favor or select only "undesirable" persons for risky research. Social justice requires that a distinction be drawn between classes of subjects that ought, and ought not, to participate in any particular kind of research, based on the ability of members of that class to bear burdens and on the appropriateness of placing further burdens on already burdened persons.[5]

The National Commission had two concerns regarding participation in research. First, it was troubled about problems of mismatch between those who participated in research and those who received beneficial therapies. In the National Commission's view, benefits should not be distributed on the basis of ability to pay, and "research should not unduly involve persons from groups unlikely to be among the beneficiaries of subsequent applications of the research."[6]

Eliminating mismatch, however, may reduce, but not necessarily resolve, concerns about research participation of individual subjects who are members of vulnerable groups. Even where individuals participate in research that promises to benefit a group or class to which they belong, they may be exposed to risk in circumstances that are not likely to benefit them directly. Society must make a difficult choice. It can encourage exclusion of subjects who are members of vulnerable groups to protect them as individuals, but by so doing increase the possibility that other members of the group will be harmed randomly by lack of appropriate therapies or inadequately investigated therapies. Alternatively, research projects can include members of vulnerable groups and expose them to risks in order to improve the well-being of the group to which they belong. The National Commission did not directly address this dilemma.

Second, the National Commission worried that some groups of subjects were especially vulnerable in research. In its view, injustice in selection of subjects occurs when some classes, such as welfare patients, particular racial and ethnic minorities, or persons confined to institutions, are "systematically selected simply because of their easy availability, compromised position or their manipulability, rather than for reasons directly related to the problem being studied."[7] Significantly, members of these classes can participate in research if availability is not the *sole* reason for selection or if selection is related to the problem being studied. Thus, social justice requires an order of preference in the selection of classes of subjects and consideration of whether some classes should be included at all, and if so, perhaps included only under certain conditions.[8]

The *Belmont Report*, however, does not describe the vulnerabilities of specific persons or groups or recognize that individuals may be a member of more than one vulnerable group. Instead, the report lumped together the classes of subjects viewed as vulnerable. Through close examination of the groups that were listed as vulnerable, it is possible to infer that the National Commission considered persons and groups vulnerable in at least two senses (obviously, some persons and groups, such as the very sick or prisoners, may be vulnerable in both senses). First, some groups (e.g., the institutionalized mentally infirm or prisoners) were included because of concern about the quality of their consent. To the extent that vulnerability is linked with questions of diminished capacity to consent, inability to protect oneself through the consent process, or inability to give informed consent, concern for selection of subjects overlaps with the principle of respect for persons. For such subjects, concern about vulnerability can be alleviated in part by efforts to improve their ability to protect themselves through imposition of additional requirements in the consent process. For example, allowing young children to participate in research might be justified by preventing mismatch between them and the class to be benefited by the research and by imposing requirements on the consent process to substitute for their lack of capacity to give voluntary informed consent.

All concerns about consent, however, cannot be alleviated by modifying the consent process or the interaction between institution or researcher and subject. For example, it may be necessary to improve conditions in prisons before permitting prisoners to participate in research. Such changes may be beyond the reach of the research establishment. Indeed, the National Commission recognized that injustice in the selection of subjects might be beyond the power of researchers or institutions to address because it "arises from social, racial, sexual and cultural biases institutionalized in society."[9]

Second, some groups were listed (e.g., racial minorities and the poor) out of concern that they would be exploited because of prejudice, unconscious bias, or relative lack of social power. Subjects that are vulnerable in this sense are capable of giving competent, voluntary informed consent. The *Belmont Report* hints at what puts them at risk when it states that injustice "arises from social, racial, sexual and cultural biases institutionalized in the society."[10] Yet, because it does not explore in depth the links between these external forces and vulnerable status, the *Belmont Report* does not suggest ways to alleviate such vulnerability.

A fair interpretation of the *Belmont Report* is that subjects should only be included in research if concerns about mismatch and vulnerability can be adequately addressed. If these concerns cannot be addressed, then subjects who are members of vulnerable groups should be excluded. Indeed, the National Commission recommended a moratorium on research involving prisoners pending the adoption of some standards. Thus, the National Commission was only willing to move cautiously in the direction of permitting individuals who were members of vulnerable groups to assume risks of research participation.

Beyond *Belmont*

In the years following the issuance of the *Belmont Report*, there has been a striking shift from the emphasis on exclusion of members of vulnerable groups to one of inclusion in research. It became clear that excluding persons from research participation in order to protect them might also cause harm. Linked with this shift was a marked change in public perceptions of clinical research. Research once regarded as risky and burdensome was now seen as an activity that offered potential benefits. Less obvious, but also of significance, was the fact that the health care crises that drove these changes also served to highlight two important trends: the increasing importance of groups in thinking about the benefits of research and the relevance of aspects of justice beyond distributive justice in trying to achieve fair inclusion in research.

BENEFITS FOR INDIVIDUALS

The 1980s and 1990s brought calls for expanded inclusion in research that involved vulnerable subjects. For example, in a little noticed effort, the Department of Health and Human Services in 1984 established the Task Force on Black and Minority Health to examine health issues of blacks and other minorities.[11] The task force's report issued in 1985 marks a shift in the federal government's concerns about black health status. Significantly, in light of the Tuskegee study and its aftermath, the report called for greater inclusion of racial and ethnic minorities in medical research. The report noted significant gaps in knowledge about the health status of African Americans and other minorities and stated that "ongoing research, particularly basic research already conducted through DHHS, applies to all populations including minorities."[12] It was primarily the emergence of HIV/AIDS in the early 1980s, however, that resulted in a more widespread appreciation of and desire for the benefits of research.[13]

The problem of HIV/AIDS was first recognized in 1981. It was not until 1986, however, that clinical trials indicated that zidovudine or AZT was efficacious in treating a condition that invariably resulted in death. As word of the research results spread, persons with HIV/AIDS clamored for access to clinical trials that tested promising therapies; frequently, participation in clinical trials was the only means of obtaining access to promising therapies.

Initially, HIV/AIDS seemed to involve only one group, gay men, who were already organized. This group was inclined to distrust the health care and research establishments so much so that in many cities, for example, gay men had organized their own clinics especially for the treatment of sexually transmitted diseases. Gay activists saw access to clinical research as important to their individual health and well-being. Persons wanted to make their own choices, including accepting the risk of unproven therapies. Risks were accepted in the hope that there might be some way of stopping or slowing down the effects of a disease that was otherwise fatal. Thus, gay activists were prepared to challenge fundamental assumptions "from traditional public health control measures to how research ought to be conducted to determining who should make decisions about access to experimental therapies."[14]

As the epidemic continued to spread, it soon became apparent that it extended beyond the gay community. Women of childbearing age, drug users, prostitutes, and prisoners also desired access to clinical trials. Access to clinical trials — especially for members of these groups — was often difficult. Many infected persons did not have access to the health care system in a way that could then steer them into research. Others were geographically isolated and not near centers of research. In addition, many of these persons were members of racial and ethnic groups or disadvantaged economic groups. Despite their membership in vulnerable groups, it seemed less important to protect such persons from the risks of research than to include them, where the only hope for survival lay in access to promising therapies often available only in research. Indeed, it seemed unjust to exclude such persons from the only avenue that offered hope of survival.

Justice in the context of the AIDS epidemic might be viewed differently now, in that there is less concern with the risks and burdens of research and more attention to fair distribution of potential benefits. Ruth Macklin and Gerald Friedland early on called attention to the new challenge:

> This problem is already apparent in the phase II study of AZT, which was performed almost entirely on homosexual male patients, excluding intravenous drug abusers. Although drug abusers were excluded according to a "medical" rationale — this group tends to be "unreliable" and "noncompliant," and hence is not a good study population — the resulting distribution of benefits was nonetheless unjust.[15]

This shift in emphasis in understanding the principle of justice that calls for fair distribution of potential research benefits has taken hold. At the close of the twentieth century, there was increased latitude given to demands by very sick patients to make their own judgments about participation in research.[16] The Council on Ethical and Judicial Affairs of the American Medical Association (AMA), in its report titled "Subject Selection for Clinical Trials," identifies three categories of potential benefits for sick patients: direct therapeutic, indirect therapeutic, and altruistic.[17] As the AMA points out, all research is designed to yield generalizable knowledge and to benefit the class from which subjects are drawn. While not designed to provide therapeutic benefits to subjects, in some circumstances subjects can derive collateral benefits from participation in research.

Although the probabilities are small, subjects may directly benefit from research participation. Subjects may not necessarily appreciate, however, the fact that research is not for their benefit. Many suffer from the so-called therapeutic misconception in which subjects assume that their interests are being advanced despite information to the contrary. Inability to distinguish between research and therapy and unrealistic hope about their conditions may make sick patients, especially those suffering from chronic, life-threatening disorders, susceptible to coercion.

By participating in research, subjects may also receive indirect therapeutic benefits such as better health care and access to social services. These benefits, however, should be viewed with caution. The prospect of such benefits exerts on persons coercive pressure to participate in research. Extreme caution is in order when one attempts to justify broader inclusion in research on inequities in access and delivery of services in the health care system.

Finally, individual subjects may receive therapeutic benefit from research participation because of the prospect of benefiting others, especially fellow sufferers. Permitting research participation for these reasons may be evidence of respect for these subjects as persons. While there is support for the view that many subjects participate in research for altruistic reasons,[18] it is not easy in practice to distinguish those who volunteer for such reasons from those who seek access to research because they believe that they will be helped. Clearly, however, true volunteers should not be prevented from trying to help others.

BENEFITING THE GROUP

Another lesson emerged from the HIV/AIDS epidemic. In addition to the fact that individuals desired access to clinical trials of possible therapies, it became increasingly clear that lack of participation in research by members of all affected population groups uncovered the possibility that research results might not be broadly applicable. For example, the participation of African Americans in clinical AIDS trials was disproportionately small in comparison to the numbers of African Americans infected with the HIV virus. Because of the possibility that African Americans might respond differently to drugs being developed and tested to combat AIDS, those concerned about the care and treatment of AIDS in the African American community called for greater participation by African Americans in these trials.[19] Moreover, the issue was of importance to many populations including women, children, and intravenous drug users. In order to achieve the goal of improving well-being of all populations, all population groups affected had to be represented in clinical trials.

The issue of including representative members of populations affected in clinical trials has resulted in important changes in research policy in the United States. For example, current National Institutes of Health (NIH) guidelines require that women and minorities be included in government-funded research. Measures have been taken to ensure that there is sufficient inclusion of minorities and women so that valid analyses of differences in intervention effects can be made. Ensuring that research results are broadly applicable, however, poses additional responsibilities for Institutional Review Boards (IRBs) that may not be constituted in a way to handle such responsibilities effectively.[20]

Ensuring that advances in science and medicine benefit all people, irrespective of gender, race, ethnicity, or age, is an important goal. Although it is possible to understand the issue of applicability of research results as a matter of beneficence because there is concern for the well-being of individuals in particular groups, this issue is properly understood as a question of justice.

Furthering well-being rests on fair access to research for all affected groups through participation of their members. This concern is reflected in the *Belmont Report*'s formulation of justice, in which it states that there should be a match between persons who participate in research and those who are the beneficiaries of the research. In this respect, I agree with Madison Powers, who writes that "[a] reasonable extension of the distributive principles of the *Belmont Report* requires a concern for the well being of disadvantaged or powerless groups as well as safeguards against the abuse of particular persons."[21]

The requirement that research results broadly apply to all affected populations rests in part on the assumption that there may be biological or genetic differences among population groups that might make it inadvisable to generalize research results to all population groups based on research with white male subjects. If this assumption has merit, then some groups may be disproportionately impacted by some diseases or conditions, or in the case of women, may have different health concerns altogether. Moreover, there is the possibility that the health needs of some groups have been ignored historically.

Inclusive policies, however, raise the specter of historical abuses in human subjects research. Although we are in the midst of efforts to distribute more widely the benefits of research and to avoid mismatch between those who participate and those who benefit, focusing on particular groups poses issues. Some argue that there are dangers that result from focusing on groups at all. As an example, the Nazis thought mainly in terms of nations and races, not individuals. Targeting a group, however, may be necessary to obtain important information or to increase resources that might improve the health status of the group. On the other hand, there is danger that conscious attention to the health needs of groups risks feeding and nourishing the stereotypes and prejudices that have historically oppressed and stigmatized groups, making them vulnerable in the first place. Promoting participation of members of vulnerable groups in research requires attending to groups, but taking care to employ policies and procedures that maximize the likelihood of achieving this goal without backlash. Attending to groups may thus require that we investigate and understand the particularities of the groups' experiences in order to learn from them. The historical experience of African Americans with research and medicine explains their reluctance to participate in research.[22] Attention should also be paid to the experiences of group members, because their insights might be helpful in constructing fair institutional and public policies. Moreover, as feminists in particular remind us, there is great need to pay special attention to the power relations of persons who are involved in or affected by practices or policies under consideration.[23] Feminists have often emphasized the oppressive practices of medical organizations and medical professions. For example, Iris Marion Young notes that the "powerless lack the authority, status, and sense of self that professionals tend to have."[24] We need to understand better the nature of these relationships if we are to protect persons in research. Subjects may need to have some influence beyond just being a subject, and we need to be creative about ways to mitigate the power imbalance in relationships between researchers and subjects.

The reasons underlying the vulnerability of certain groups — "social, racial, sexual and cultural biases institutionalized in society" — are the same today as in previous decades. [25] There are still examples of research in which persons suffer physical or dignitary harms. For example, in a recent government-sponsored measles vaccine study in which a large population of the subjects was African American and another minority, parents were not informed that the vaccine was experimental and not licensed for use in the United States.[26]

In some circumstances, we may have greater cause for worry today than in the past. For example, one of the pronounced changes in the last decades has been increased willingness of individuals to participate in therapeutic research. The economically disadvantaged, especially in view of dramatic changes in the health care system, may be more likely to be research subjects because such participation may provide them with access to health care benefits that they otherwise might not be able to obtain. Their need for health care may make them susceptible to exploitation and coercion. Adding protections to the informed-consent process, particularly in view of the so-called therapeutic misconception discussed earlier, may be insufficient to protect them.

The issue of reordering research priorities or making changes in design or methodology indirectly implicates selection of subjects in that members of vulnerable groups must be included in clinical research if desired benefits are to be obtained. Yet this issue takes us further. First, it is clear that modifying or adjusting research strategies, priorities, or resources is beyond the competence of IRBs. Such decisions are made in the heart of the research establishment outside of the interactions between researcher and subject. A second inference is that justice may require more than securing greater inclusion of women, minorities, and other groups in research to derive the benefits of research.[27] The objective is for groups to play a more central role, not just in the deliberations of IRBs but in the entire research process, including a role in setting the research agenda and determining what counts as beneficial research.

The last several decades make clear that fairness requires that we focus on more than the distributive features of justice — specifically compensatory justice and procedural justice — that were ignored in the *Belmont Report*. The issue of broadening power and influence in all areas of the research establishment goes beyond distributive concerns to matters of decision making and modification of institutional arrangements that have long been in place. This is the realm of procedural justice that "applies to a wide variety of social, legal and institutional matters in which achieving a fair or unbiased result is dependent on adherence to a set of well-ordered procedures."[28] For example, including women, African Americans, or others in a process to establish research agendas can be understood as procedural justice. Another example might be involvement of African Americans in a process for clarifying the use of race as a variable in epidemiological or medical research.

Compensatory justice takes us beyond distributive justice in the sense that it addresses redress for harm or wrongs. This remains an important issue, as some subjects are harmed through participation in research. An example of this is making monetary payments to Tuskegee survivors for the harms suffered in the Tuskegee study. The Advisory Committee on Human Radiation Experiments took compensatory justice into account when it recommended that the federal government consider compensating research subjects who suffered physical injury or dignitary harm as a result of participating in government-sponsored research. Compensatory justice may also play an important role in redressing injury to groups. Remedial measures may be required where it can be shown that specific health interests of certain groups have not received fair allocation of research resources or where policies caused harms. For example, harm may result from failure to study diseases that disproportionately afflict specific groups. Harm may also result from inadequate selection policies. These practices can be corrected by attending to selection procedures and instituting a fair standard for allocating resources. Importantly, however, mere reversal of past policies will not completely alleviate racial and ethnic disparities in health created by historical injustice. For example, race-neutral or gender-neutral allocation policies will prevent the gaps between black and white or male and female from widening, but they will not close the gap. Thus, affirmative efforts to compensate for historical practices may be required.

Conclusion

The *Belmont Report* emphasizes that some groups in the population are vulnerable and should be protected from coercion and exploitation. Events of the last thirty years demonstrate that too much protection, especially in the form of exclusion from research participation, can raise questions of unfairness as well. This state of affairs raises questions of how the requirements of justice are to be understood in the future.

The *Belmont Report* requires fair distribution of both burdens and benefits. As a principle it remains applicable today and into the future. What is required in the selection of subjects is both to create opportunities for inclusion and to protect vulnerable subjects from exploitation and coercion. This balancing will require attention to the role of groups as well as individuals. Creation of opportunities for inclusion may require us to look beyond the researcher-subject interaction to other parts of the research enterprise. Protection of subjects might require us to branch out into the realm of health care or at least to understand better the relationships between health care and research. Finally, it may be necessary to consider more than just the distributive aspects of justice to ensure fair inclusion of subjects in research.

Since the *Belmont Report*, events have demonstrated that research does provide benefits and that it is possible to distribute those benefits more broadly over the population. Providing research benefits to broader segments of the population is in progress, but this effort is in its infancy, and we should proceed cautiously. Even efforts to provide benefits may backfire because of unconscious bias or reliance on negative stereotypes about groups. For example, attempts to address sickle cell anemia in the African American community in the 1970s remind us that well-intentioned efforts may cause harm.[29]

Including vulnerable groups in research will require greater attention to the vulnerabilities of *each* group to achieve optimal levels of participation. This requires understanding the reasons for vulnerability and then devising institutional changes to address the issue. Recently issued regulations allowing waiver of informed consent from research subjects or their surrogates in certain cases of emergency room research have potential for benefit, but they may also pose risks to minorities.[30]

In short, as we move into the twenty-first century, justice requires that we continue to protect subjects of research as well as work to include them more extensively in research. As Madison Powers notes, "Just because the protective conception may have been exaggerated in some instances and thus operated to the detriment of some, it does not mean that the protective conception ought to be abandoned."[31]

Notes

1. Carol Levine, "Changing Views of Justice after Belmont: AIDS and the Inclusion of 'Vulnerable Subjects,'" in *The Ethics of Research Involving Human Subjects: Facing the Twenty-first Century*, ed. Harold Y. Vanderpool, 105–26 (Frederick, MD: University Publishing Group, 1996).

2. Albert R. Jonsen, *The Birth of Bioethics* (New York: Oxford University Press, 1998), 413.

3. U.S. Department of Health, Education, and Welfare, National Commission for the Protection of Human Subjects of Biomedical and Behavioral Research, Ethical Principles and Guidelines for the Protection of Human Subjects, *The Belmont Report* (Washington, DC: Government Printing Office, 1979), 8 [hereafter *Belmont Report*].

4. Ibid., 9.

5. Ibid., 18.

6. Ibid., 10.

7. Ibid., 9–10.

8. Ibid., 18.

9. Ibid., 19.

10. Ibid.

11. U.S. Department of Health and Human Services, Task Force on Black and Minority Health, *Report of the Secretary's Task Force on Black and Minority Health* (Washington, DC: Government Printing Office, 1985), 7.

12. Ibid.

13. Levine, "Changing Views of Justice after Belmont," and Charles Weijer, "Evolving Ethical Issues in the Selection of Subjects for Clinical Research," *Cambridge Quarterly of Healthcare Ethics* 5 (1996): 334–45.

14. Jeffrey Levi, "Unproven AIDS Therapies: The Food and Drug Administration and DDI," in *Biomedical Politics*, ed. Kathi E. Hanna, 9–37 (Washington, DC: National Academy Press, 1991).

15. Ruth Macklin and Gerald Friedland, "AIDS Research: The Ethics of Clinical Trials," *Law, Medicine & Health Care* 14 (1986): 273–80.

16. Baruch Brody, "Research on the Vulnerable Sick," in *Beyond Consent*, ed. Jeffrey P. Kahn et al., 32–46 (New York: Oxford University Press, 1998).

17. Council on Ethical and Judicial Affairs of the American Medical Association, "Subject Selection for Clinical Trials," *IRB* 20, nos. 2–3 (1998): 12–15.

18. U.S. Department of Energy, Advisory Committee on Human Radiation Experiments, *Final Report: Part III* (Washington, DC: Government Printing Office, 1995).

19. Walfaa El-Sadr and Linnea Capps, "The Challenge of Minority Recruitment in Clinical Trials for AIDS," *Journal of the American Medical Association* 267 (1992): 954–57.

20. Weijer, "Evolving Ethical Issues in Selection of Subjects for Clinical Research."

21. Madison Powers, "Theories of Justice in the Context of Research," in *Beyond Consent*, ed. Jeffrey P. Kahn et al., 147–65 (New York: Oxford University Press, 1998).

22. Vanessa N. Gamble, "A Legacy of Distrust: African Americans and Medical Research," *American Journal of Preventive Medicine* 6 (1993): S35–S38.

23. Susan Sherwin, "Feminism and Bioethics," in *Feminism and Bioethics*, ed. Susan M. Wolf, 47–66 (New York: Oxford University Press, 1996).

24. Iris Marion Young, *Justice and the Politics of Difference* (Princeton, NJ: Princeton University Press, 1990), 57.

25. *Belmont Report*, xx.

26. Charles Marwick, "Questions Raised about Measles Vaccine Trial," *Journal of the American Medical Association* 276 (1996): 1288–89.

27. Patricia A. King, "Race, Justice, and Research," in *Beyond Consent*, ed. Jeffrey P. Kahn et al., 88–110 (New York: Oxford University Press, 1998).

28. Anna C. Mastroianni, Ruth Faden, and Daniel Federman, *Women and Health Research: Ethical and Legal Issues of Including Women in Research*, vol. 1 (Washington, DC: National Academy Press, 1994), 76.

29. King, "Race, Justice, and Research."

30. Annette Dula, "Bearing the Brunt of the New Regulations: Minority Populations," *Hastings Center Report* 27 (1997): 11–12.

31. Powers, "Theories of Justice in the Context of Research," 161.

11 *Belmont* Revisited through a Feminist Lens

Susan Sherwin

One of the most striking things about the *Belmont Report*, looking back thirty years later, is how incisive and comprehensive it manages to be in so few pages. Even though it appeared several years before explicitly feminist critiques of research ethics, it anticipated many of the concerns that have troubled feminists in subsequent years. Unfortunately, however, the issues that concern feminists are, at best, merely mentioned in the midst of many other recommendations, with no elaboration or explanation and are therefore easily overlooked. Other important issues are totally absent from the report. Moreover, while several statements are open to feminist readings, it is not clear that feminist interpretations of these passages were consciously intended. Given this ambiguity and the lack of explicit attention to feminist analyses, I doubt that anyone not already persuaded by feminist critiques would be disposed to interpret the recommendations of the report in accordance with feminist commitments. In practice, few of the cryptic references that can be inferred to support feminist concerns have been emphasized or taken up by those who rely on the *Belmont Report* as a basis for research ethics.

I propose, therefore, to explore what the *Belmont Report* might look like if approached from a distinctly feminist perspective.[1] This strategy will present how the report did deal with these matters and reflect on the effectiveness of its recommendations for meeting feminist concerns in the practice of research. It will also help to ensure that the issues and analyses that emerge from a feminist perspective can be made more explicit and compelling in the future. By exploring ways in which the *Belmont Report* comes close to articulating feminist concerns without actually addressing them, we can gain a better sense of what is needed to develop revised research ethics guidelines that more fully capture the demands of social justice.

I base my analysis on the set of issues identified by the Canadian Feminist Health Care Ethics Research Network (hereafter the Network), an interdisciplinary, collaborative research network of eleven scholars and practitioners that I coordinated under the sponsorship of a Strategic Research Network grant from the Social Sciences and Humanities Research Council of Canada from 1993 to 1998.[2] Among other projects, the Network reflected on ways of formulating ethics guidelines for research involving humans that would be consistent with our collective understanding of feminist ethics. We organized our interventions around four themes that we hoped to see incorporated in the final report of the Canadian Tri-Council Working Group, established with a mandate to develop guidelines for Canadian federally funded research involving humans. I shall use these four themes to guide my discussion of the *Belmont Report*.[3]

Research, Therapy, and Innovative Practice

The *Belmont Report* addresses two of the Network's concerns directly, and the other two indirectly, though never quite in the spirit we would like. One theme that is directly addressed concerns the importance of clarifying the distinction between research and therapeutic practice, with particular attention to the ambiguous status of innovative practice. In the past thirty years it has become apparent that precise distinctions between research and therapy are impossible. Nonetheless, Network members believe that it is essential to distinguish innovative practice from both routine therapy and formal research. Often, innovative therapy is pursued without any clear demarcation as to its experimental nature and without the careful attention to ethics that is appropriate to experimental efforts. The *Belmont Report* makes explicit the fact that clinicians sometimes depart from standard practice in a fashion typically identified as experimental practice.

Feminists take a particular interest in this subject because we are attentive to many of the ways in which established medical norms have resulted in harms to women. Most feminists are aware of the hazards associated with the long — and continuing — history of innovative practice in many areas of reproductive medicine that are directed solely at women. The harms that resulted from the widespread use of DES[4] and thalidomide[5] in pregnancy provide clear warnings of the danger that can ensue when therapies become common practice prior to thorough testing. Despite the documented dangers of unregulated innovative practice, such approaches are common in many clinics involved in assisted reproduction. For example, clinics often modify doses of fertility drugs according to their own intuitions and agendas, even though such actions contribute to increased rates of multiple births and other serious side effects. The recent long-term study of widespread use of hormone-replacement therapy[6] for menopausal women in the absence of supporting scientific evidence of its safety and efficacy is yet another example of widespread medical practices targeted at women that become entrenched without adequate review.

Although the *Belmont Report* does acknowledge the existence of innovative practice, its interest in this phenomenon is not the same as that of the

Network. In fact, it seems that the report's authors had very different goals in mind when they distinguished innovative therapy from research programs. The main thrust of their discussion is to excuse clinicians engaged in innovative therapeutic efforts from the need to comply with the rigorous requirements of full-blown research projects. By treating innovative therapy simply as well-intentioned efforts to improve practice, the *Belmont Report* essentially exempts all but "radically new procedures" from the demands of external review. Even then there is only a demand that formal research studies be undertaken "at an early stage" of development of radically new procedures. No explanation is provided for determining what constitutes radically new therapy and what counts as merely an acceptable modification of existing practice. There is no guidance, for example, as to where to fit the practice of increasing the dosage of fertility drugs beyond manufacturers' recommendations. Should it be viewed as a case of radically new therapy, an innovative attempt to increase success rates, or a dangerous malpractice? And who is to be the judge of this: the clinicians, medical practice committees, or research ethics boards? Without clear instruction as to when clinicians are required to send modified procedures for external review, there is little likelihood of uniform compliance with this requirement.

In its haste to distinguish innovative therapy from formal research, the *Belmont Report* fails to be explicit about the ways in which the former also differs from proven therapies. The drafters seem to be confident that not all innovations require the same procedural steps of external review as do full-fledged research protocols, and they are probably correct about this. Nonetheless, it is essential to explain what norms are appropriate to innovative therapy, because there is clearly a need for ethics guidelines to govern innovative procedures. For instance, it should be specified that informed-consent requirements comply with the standards applied to research efforts, rather than the looser standards associated with established therapy. Patients who participate in innovative practice should be informed of the experimental nature of their therapy and of the ways in which their particular course of treatment varies from established practice. They should know the success and failure rates of related efforts and the possibility of added risk associated with the new interventions. In sum, research ethics guidelines need to make clear the distinction among therapy, research, and innovative practice, and they should insist that norms be developed for the conduct of innovative practice.

Moreover, a feminist lens helps us to appreciate the importance of reflecting on the diverse ways that members of differently situated social groups might be involved in unregulated innovative therapy. Specifically, it reveals that members of oppressed or disadvantaged groups can be at particular risk of being unwittingly recruited into innovative therapies. When clinicians are experimenting with new techniques, they may feel fewer constraints if the experimental intervention is to be used primarily on members of socially devalued groups (e.g., women, poor people, racial minorities, or people with disabilities). Therefore, when an innovative procedure is primarily aimed at socially vulnerable populations, it is especially important to ensure that the associated risks are reasonable and that the anticipated benefits are honestly reported.

Selection of Research Subjects: Risks of Overrepresentation

This brings me to the general question of selection of research subjects, a topic that receives considerable attention in the *Belmont Report*. In our own deliberations, the Network came to the conclusion that this topic really involves two questions: we perceived both a danger of exploitation through overrepresentation of certain disadvantaged groups in some studies and a danger of underrepresentation of disadvantaged groups in other areas of research. However, the Report either conflates these questions or it ignores the second (depending on how charitably we read it). Moreover, even though the drafters of the *Belmont Report* did pay considerable attention to the first question, their concerns seem to be far more limited than feminists would like.

Quite rightly, the *Belmont Report* devotes considerable space to discussion of the first question. Exploitation of vulnerable groups as research subjects is a central issue in most discussions of research ethics. Indeed, many credit the birth of the modern bioethics movement to the public outrage that followed revelations of horrifying abuse of vulnerable populations for the sake of scientific research. The postwar Nuremberg war crimes trials exposed terrible tales of inhuman treatment of concentration camp victims in support of a questionable medical research agenda. This knowledge led to the Nuremberg code of research ethics, the first systematic effort to effect explicit ethical constraints governing the conduct of research involving human subjects.

Public concern about research ethics gained new life in the early 1970s when the American media learned of the ongoing Tuskegee syphilis study (see Patricia King's chapter in this volume). In this case, disadvantaged, poorly educated black men were used, without their knowledge or consent, to study the untreated course of a well-known — and by then treatable — disease. This was not an isolated case. Institutionalized mentally retarded children were used for hepatitis studies at Willowbrook, cognitively impaired elderly men were used for an unrelated cancer study at the Jewish Chronic Disease Hospital in New York, and nonconsenting, poor Mexican American and Puerto Rican women were used for early tests of the birth control pill. These cases make clear that even in peacetime America researchers could lose sight of the basic ethical commitments of respecting human dignity, avoiding unnecessary harm, and attending to the demands of justice. Widely documented in the media and professional literature, these cases formed the backdrop to the work of the team that drafted the *Belmont Report*.

Indeed, the section on justice makes reference to the fact that "the burdens of serving as research subjects fell largely upon poor ward patients," and it cites the "flagrant injustice" of the Nazi exploitation of unwilling prisoners and the Tuskegee syphilis study as evidence of the shameful history of research abuses. The *Belmont Report* warns us to be wary of unfairly exploiting certain "classes" by overinclusion of members from particular groups in studies because of their "easy availability, their compromised position, or their manipulability," for example, "welfare patients, particularly racial and ethnic minorities, or persons confined to institutions." These are important constraints, and it is very helpful that the *Belmont Report* recognizes that whole groups of people — not just isolated individuals — can be at particular risk of exploitation in research contexts.

Moreover, these concerns are reiterated and elaborated in the concluding section of the *Belmont Report*, which deals with the selection of research subjects. Once again, the reader is reminded that "the principle of justice gives rise to moral requirements that there be fair procedures and outcomes in the selection of research subjects." Anticipating feminist concerns, the report explicitly states that justice is relevant to the selection of subjects at the social as well as the individual level, and it quite properly warns against the selection of only "undesirable" persons for risky research (quotes in the original *Belmont Report*). It observes that potential subjects belong to "classes" and warns against selecting subjects from groups that may have less ability to bear the burden of research or who are already bearing an excessive burden. The report goes so far as to observe that "injustice arises from the social, racial, sexual and cultural biases institutionalized in society." These are welcome words to a feminist who is deeply concerned about ways that social practices may affect existing patterns of oppression.

No sooner does the *Belmont Report* note that injustice affects social groups in many complicated ways than it reverts back to warning that it is easy accessibility that exposes groups (or classes) to high risk of exploitation. In so doing, it implies that the major risk is that of accessibility, noting that this may be a product of oppression (or injustice), but it fails to observe other features that may make oppressed groups overly attractive to researchers. Hence, while feminists can applaud the report's demands for scrutiny of subject selection strategies and its recognition that whole "classes" or groups of people may be at particular risk by virtue of their easy accessibility to researchers, we cannot be satisfied with its formulations of the problem. Significantly more must be said about the risks of exploitation that are associated with membership in an oppressed social group.

What is missing is an analysis of oppression that explains how being a member of an oppressed group in society makes one particularly vulnerable to exploitation in research contexts. Such an account would explain why it is that membership in an oppressed group increases one's "accessibility" to researchers. This requires discussion of the fact that members of oppressed groups are (generally speaking) economically as well as socially disadvantaged. As a result, they are more likely than others to be dependent upon publicly funded health care services; in the United States, this means different treatment and greater accessibility to researchers than patients who can obtain private health care.

This is not the only difference between oppressed and privileged groups, however. Oppressed groups are generally regarded as being of lesser importance to society than members of more powerful and privileged groups. Their lives are frequently devalued and their interests overlooked in public policy efforts. It is not so much that they are easily "manipulable" by virtue of their own diminished competence (for, often, their competence is perfectly adequate), but that they lack the political clout to be treated as fully equal members of the community. Society accepts harms to oppressed groups, but not to members of more privileged groups. Thus, it is not just that the black men of Tuskegee were easily manipulated, it is also that society did not think their lives important enough to ensure proper medical treatment for them once it became available. Nor was the "easy availability" of concentration camp inmates sufficient to explain the cruel research program of the Nazis. We must recognize that these offensive research programs were conducted within cultures that did not grant full humanity to members of the groups selected for research use.

The real problem at the heart of these and other classic cases of research abuses is the ways in which researchers are inclined to accept prevailing cultural values that view some social groups as inferior to other groups in society. The dominant culture of anti-Semitism in Nazi Germany was an essential element of the Nazi research program, as was the culture of racism in the American South over the years of the Tuskegee study. Similarly, discriminatory attitudes toward people with mental handicaps, elderly people with dementia, and poor Hispanic women were important factors in the Willowbrook hepatitis study, the Jewish Chronic Disease Hospital cancer study, and early birth control trials, respectively.

As Susan Wolf[7] has argued, difference associated with membership in an oppressed group has been at the center of virtually all the cases that have generated efforts to develop and refine policies of research ethics. She argues that

> bioethics was born of outrage at scandals in which difference figured large. . . . The field has come of age decrying the wrongs of physicians and scientists enacting the prejudices of their day, whether those professionals were extolling racist eugenics or excluding women from research trials.[8]

Wolf notes that even though membership in disadvantaged social groups has been central to the mistreatment of victims in the most disturbing cases of research abuse, this feature has been largely excluded from bioethical analysis. Bioethicists generally, and the drafters of the *Belmont Report* specifically, seem to assume that treating individuals with the procedural demands of individual respect (primarily by obtaining their informed consent) will resolve the disadvantages of membership in a vulnerable social group. They fail to recognize the multiple ways in which group membership affects individuals' potential roles in research trials.

In contrast, when we use a feminist lens to examine the problem of group vulnerability, we begin with an analysis of the nature of oppression and its differential impact on differently situated social groups. This leads us to frame the problem differently. A feminist lens reveals that many research abuses are not just unhappy experiences of assorted individuals who happen to share certain characteristics. It makes visible the many ways in which membership in an oppressed group increases a person's risk of being exploited in research. Feminism reminds us that research is conducted in a particular cultural climate and that researchers are likely to share the social biases of their day. In other words, a feminist lens looks at the context as well as the details of each research project.

All patients are "accessible" to clinical researchers in that they are actively engaged with the health care system; if they are hospitalized, it is especially easy to gain access to them. What needs explanation is why it is that welfare patients are at so much higher risk than private patients of being exploited in research programs and why membership in a racial or ethnic minority can exacerbate the problem. This task requires a discussion of oppression and the recognition that oppression extends far beyond the "classes" identified in the *Belmont Report*. In particular, it demands attention to the fact that those who are oppressed are also devalued by society, because it is their devalued status that makes it difficult for others to recognize some forms of treatment of such people as abusive.[9]

To truly protect against exploitation and abuse of subjects from vulnerable groups, therefore, it is important to conduct a more thorough analysis of ways in which membership in different sorts of groups places individuals at risk. The *Belmont Report* was sensitive to the ways in which ready accessibility (e.g., institutionalized patients), dependence (e.g., poor people in need of free health services), and explicitly coercive environments (e.g., prisons) could put some groups at risk by reducing members' sense of freedom to refuse participation. But it did not acknowledge the dangers that arise when society and its researchers view members of oppressed groups as being less valuable, less competent, and/or less likely to be interested in the details of the research than those who belong to more powerful groups.

Selection of Research Subjects:
Risks of Exclusion and Underrepresentation

Moreover, the question of subject selection becomes even more complicated when we note that there is another side to the selection problem. In addition to the dangers of exploiting members of oppressed groups by overrepresentation in research trials, there is also the companion worry that sometimes exclusion or underrepresentation in trials can also be unjust to members of oppressed groups. While a general principle of justice can perhaps be presumed to include both problems, it is necessary to make explicit the distinctions between them in order to ensure that both mistakes are avoided.

At the time the *Belmont Report* was written, the serious harms associated with objectionable research programs seemed the most urgent ethical issue. Attention was focused on the risks and harms associated with participation in a research trial, and few people were thinking of the opportunities sometimes associated with being a research subject. It is now quite clear, however, that subjects often stand to benefit from participation in research trials. For example, membership in a research protocol can provide subjects with the sole or best chance of gaining access to needed therapy. This may be because they live in a country (e.g., the United States) where the distribution of health services is profoundly unequal and participation in a research trial gives economically disadvantaged patients access to high-quality health care that would otherwise be unavailable to them. Or it may be that they suffer from a condition in which there is no effective therapy currently available such that access to an experimental treatment may represent their only hope for relief or even survival. In the face of this reality, it is unjust to exclude people on the basis of their membership in particular groups from trials that may be more beneficial to them than any available alternative. And yet, women have frequently been routinely excluded from many therapeutic trials on the basis of their sex.[10]

In addition, it is now evident that exclusion of certain groups from research trials results in knowledge gaps that threaten the interests of all members of that group. If specific efforts are not made to determine whether or not there are differential impacts of particular interventions on members of different groups, clinicians will subsequently lack the data necessary to care properly for members of the neglected or excluded groups. Hence it is ethically objectionable that many clinical trials investigating therapeutic interventions have been conducted — initially, and often solely — on young to middle-aged males. Such choices probably reflect the oppressive assumption that these men represent the norm for the species and that selecting people from "other" groups would complicate data and skew results. The idea, apparently, is to first determine how some intervention works with a "normal" population and then to extrapolate that data to members of "other" groups in the population.

As feminists Rebecca Dresser[11] and Veronica Merton[12] have eloquently argued, these assumptions of normality are unacceptable and dangerous. Not only do they help to perpetuate oppressive stereotypes by treating women, the elderly, and members of certain racial minorities as "different" rather than typical members of the species, but they also produce results that are dangerously incomplete. For if it is indeed the case that including a heterogeneous population will complicate the study and alter the results, it must mean that factors such as body size and type, lifestyle considerations, or hormonal differences can alter the effects of the intervention. If that is the case, then data from a homogeneous section of the population cannot be safely extrapolated to other groups. Without evidence that women will respond in the same way as men, there is no basis for presuming that the intervention in question (whether it is a drug, a type of surgery, an exercise program, or an alteration in diet) will affect women as it has men in the study. Hence, when physicians subsequently treat women, they must refer to far less reliable data than that available for treating men. This difference in group-specific knowledge results in inferior care for each group that has been underrepresented in the study: typically women, the elderly, people with chronic illnesses (especially mental illness), and people of racial or ethnic minorities. It is therefore a violation of justice. To guard against such injustice, it is necessary to insist formally on the inclusion of representatives of all relevant groups in sufficient numbers to allow researchers to analyze the data for each group. Such an approach would require establishing a burden of proof that researchers must provide if they seek to exclude particular groups from the pool of subjects (e.g., excluding women from studies of prostate cancer).

Given the importance granted to justice considerations in the *Belmont Report*, it is worth asking why there is so little attention paid to the problem of exclusion. The explanation seems to lie in the report's treatment of justice as a consideration among individuals alone, and not between groups. Under the section on the principle of justice, there is a statement that "equals ought to be treated equally," followed immediately by the question of who should count as "equal." To guide reflection on the equality principle, the *Belmont Report* identifies several formulations regarding the just distribution of benefits and burdens. Each principle takes an individualistic approach to the problem of equality and speaks of distributing benefits (and burdens) to individuals according to some appropriate characteristic of the person. There are, however, many ways in which membership in disadvantaged social groups affects people's eligibility for benefits and assignment of burdens; for example, women are traditionally assigned far greater responsibilities than men for child care and other aspects of domestic life. Such disproportionate arrangements are difficult to see if we consider people simply as individuals. To address all forms of injustice, we must look for patterns of difference in expectations and treatment and consider whether various sorts of differences are systematic or random variations.

Also, some benefits and burdens are inherently collective and apply primarily to groups rather than to individuals. For example, unless we recognize the ways in which different groups are affected by research designs that focus on certain subject populations and exclude others, we cannot appreciate the differential harms that face differently situated individuals. After all, it is not simply as individuals that women are disadvantaged by their exclusion from research trials; rather, it is as members of a group for which the relevant data are unavailable. Thus, we need to ascertain how different groups are treated with respect to eligibility for collective benefits. We need, that is, a feminist lens that guides us to an understanding of justice that looks beyond the role of individuals in isolation and takes into account social factors that are associated with group membership.

Setting the Research Agenda

The fourth area of concern identified by the Network involves questions about the research agenda itself. A full ethical analysis of research with humans must reflect not only on how research is conducted but also on what research topics are pursued. That is, we must consider what questions are investigated — and what questions are neglected. In particular, we ought to ask whose interests are served by the specific research projects undertaken and whose are not.

We have already seen ways in which the selection of subjects has historically favored privileged groups in society and exposed members of disadvantaged groups to significant harms. In addition, it is sometimes the case that research programs are undertaken in pursuit of knowledge that can be anticipated to favor the interests of some social groups while neglecting or actually threatening the interests of others. For example, abundant public health data demonstrate the connections between socioeconomic factors and health, linking poverty and illness. Yet the health care research agenda continues to focus on biological over social factors and on treatment over prevention. Such an agenda supports a health policy that neglects social factors in favor of physiological interventions for those who can afford them. Moreover, despite explicit contrary advice in the *Belmont Report*,[13] often the biomedical interventions that are developed involve complicated technology whose costs are beyond the reach of most people who might need them.

Consider the current excitement about the promises of biotechnology. Several research projects are aimed at developing the means to extend the human life span for another twenty to fifty years. Through use of cloning to develop replacement parts for those that wear out or become diseased, supplemented by ongoing hormone treatment to delay the onset of debilitating old age, researchers are exploring ways to ensure that those who can afford such care will live well over a century. In Peter Singer's[14] words, these researchers are in active pursuit of the "Fountain of Prolonged Middle Age." Such research goals are questionable at a time when African Americans do not have as high a life expectancy as European Americans and when many nations in the world still experience devastating famines, unacceptably high maternal and child mortality rates, and intolerable levels of HIV infection.

It is not only costs that determine whether or not a research program is likely to foster the interests of one group at the expense of another. Again, we can look to debates around biotechnology for insight into how different groups experience its "advances." Research in molecular biology has provided us with ways to identify a growing number of "deleterious" genes. This research supports prenatal testing for a long list of genetic anomalies in fetuses. But the wonders of genetics are problematic, to say the least. There is a tendency to overemphasize the power of genes and to become genetic determinists who assume that genetic traits fully determine outcomes. In the public domain, and in the practice realm of prenatal diagnosis, there is little discussion of the fact that genes interact with environments and do not by themselves "cause" anything. A genetic tendency toward most diseases reflects only a higher probability of developing that disease, not an inevitable outcome. And even where the presence of the disease can be reliably predicted, it is usually impossible to determine prenatally how severely affected a child will be. Hence, a prenatal diagnosis of Down syndrome or sickle cell disease tells us little about the life prospects of the future child, unless, of course, society is so intolerant of people with this condition that we can anticipate social factors that will make their lives miserable.

Barbara Katz Rothman[15] has documented the ways in which development of prenatal diagnosis technology has transformed the experience of pregnancy for most Western women, forcing them to confront choices that many would prefer not to have to make. This situation is the product of a research agenda whose service to women is debatable; nonetheless, now that the technology has become a routine part of obstetrical care, very few pregnant women feel free to refuse it. The routine use of prenatal testing is premised on the problematic assumption that disabilities are so tragic that it is better to prevent the birth of children with predictable disabilities than to try to adapt the environment to meet their needs. Many perceive this approach as evidence of significant prejudice against people with disabilities and, moreover, as legitimizing and deepening such prejudices by making avoidance of children born with disabilities a normal goal of medical obstetrical care.

In another area, efforts are under way to create a human gene bank by collecting tissue samples from racially and ethnically distinct groups that are threatened with extinction. These efforts have created significant anxiety and mistrust in some communities where the prevailing sentiment is that the scientific research community values their DNA far more highly than it values the people themselves. Many see this research program as simply another instance of colonial exploitation and further devaluing of people considered "other." In light of the problematic history of genetic research — including its deep roots in eugenics and genocide[16] — it is not unreasonable for minority communities to feel unease when recruited or conscripted to participate in such programs.

The *Belmont Report* does anticipate some of these worries, and it is instructive to compare its approach with the feminist recommendations I am offering. The report speaks of the need to ensure that publicly supported research "not provide advantages only to those who can afford them," and it states that "such research should not unduly involve persons from groups unlikely to be among the beneficiaries of subsequent applications of the research." These are worthy sentiments that bear repeating more than a quarter of a century later, but I would disagree about the need to restrict these constraints to publicly funded research. After all, fair distribution of benefits is a desirable goal for any project, and the responsibility to improve the health of the worst off is a social one that should be shared by all members of society. Furthermore, the boundaries between publicly and privately funded research are by no means clear or complete.[17]

Significantly, the *Belmont Report* also acknowledges that "risks and benefits may affect the individual subjects, the families of the individual subjects, and society at large (or special groups of subjects in society)." This is a clear expression of the fact that research activities may have profound effects that extend far beyond the particular harms or benefits experienced by research subjects. No sooner is the possibility of group or societal harm announced, however, than it is then ignored as the analysis again focuses on potential harms and benefits to individuals. Here too the drafters seem to have failed to grasp the implications of their own analysis. Feminist analysis demands, however, that we explore in far greater detail the ways that research programs may affect groups, especially oppressed groups, and also society at large. Without such elaboration, there is little to guide researchers or Institutional Review Boards (IRBs) regarding the need to consider social as well as individual risks and benefits of research agendas.

Even though the *Belmont Report* does identify the problem of research programs creating a disproportional impact on certain groups, it moves far too quickly to ethical analysis solely at the level of the individual. For instance, it relies heavily on the privatized practice of obtaining informed consent from fully briefed individual subjects to protect against abuse of subjects. An explicit discussion of the relationship between oppression and research would make clear why even the best efforts at obtaining informed consent cannot be sufficient to establish that the research is morally acceptable.[18] As it is, the report lacks the commitment to pursue analysis at the level of groups as well as at the level of individuals.

Adopting a feminist lens, we perceive more clearly the need to step back from particular research programs in order to consider the broad questions of how research topics are selected and framed and how research priorities are assigned. By focusing our attention on disadvantaged social groups, a feminist lens illuminates the multiple ways in which research programs tend to favor the interests of the most privileged groups in society and to neglect or threaten those of the least advantaged. It helps us to appreciate that such preferential results need not even be the result of deliberate malice or conscious self-interest. Often, it is simply a reflection of society's routine, automatic privileging of the interests of the more powerful over those who are disadvantaged by adopting the perspective of the former as the norm.

The question thus arises as to how to counter such entrenched habits. The first step is to ensure that instances of privilege and disadvantage are made visible to those making fundamental decisions about research programs. Typically, members of privileged groups are unconscious of their own privilege, and they presume that their perceptions are untainted by the advantages they take for granted. The best observers of oppression tend to come from the ranks of those who are being oppressed, as they have the most to gain from identifying and understanding the workings of oppression.

Hence, people who are oppressed often provide distinct insights into the ways in which a particular research program is likely to affect existing patterns of oppression. Frequently, they notice aspects of the research that are apt to be overlooked by those who are less vulnerable.

Awareness of oppression is not the only important issue, however. Many other relevant features may vary with the perspective of particular agents. Feminist epistemologists have explained that individuals approach problems from specific social and historical locations, and these particular backgrounds inevitably affect their perceptions.[19] The differences in experiences and interests between members of privileged groups and members of oppressed groups may lead to important differences in values and beliefs. It is therefore problematic to grant any one group the authority to make all decisions about the values and interests of society that underpin determination of a country's research priorities.

It is also a difficulty that the bodies that set research priorities are, typically, dominated by members of privileged social groups. Both the community of researchers and that of funding agencies tend to be composed of well-educated, well-connected members of society. These relatively homogeneous groups set research priorities that reflect the issues that seem most salient to them. They tend to be less sensitive to the health priorities of other groups, for their members do not share the experiences of less advantaged groups, nor are they encouraged to make time to listen to people outside the research community. Sharing a common perspective helps to reinforce their sense of the universality and legitimacy of their own priorities, and it may hide from them the need to consult other social groups about research priorities.

This is a structural, rather than a personal, failing. Those who decide about the choice and design of research projects approach their tasks wearing lenses that reflect their own social positions and histories. These lenses make some matters clear and important, but they obscure other subjects. People can learn to expand their vision by learning to listen carefully to the claims of others with different perspectives, but these borrowed lenses will always fit imperfectly. Ultimately, the best way of ensuring fairness in the setting of the research agenda and in determining the details of research design is to make certain that a wide variety of lenses is available to decision-making bodies. This requires fair representation of diverse social groups throughout the full research process.

In other words, it is not sufficient to ask research bodies to take feminist principles into account in their deliberations. They cannot capture the demands of social justice by striving to ensure that the selection of research subjects is fairly distributed among different social groups if the topics have already been chosen, the design has been determined, and the amount of funding settled. The project itself may be one that perpetuates discrimination (e.g., research into the genetic basis of homosexuality may pathologize homosexual behavior in ways that support homophobia). Moreover, unless the decision-making bodies are in a good position to recognize discrimination, they will not be able to demand social justice. For example, they must be able to appreciate the ways in which similar treatment of people differently situated may have very different effects (e.g., free medical care while undergoing an experimental treatment has different meaning for patients with and without health insurance).

The best way of promoting equality in research design, execution, and outcome is to make certain that each decision is approached from a variety of perspectives (through a variety of lenses). Because the perspectives of the least advantaged are most often overlooked, it is especially important to have representatives of particularly vulnerable groups participating in the decision making at each step.[20] Hence, including representatives of vulnerable groups on IRBs makes it far more likely that all research subjects are treated fairly. Similarly, inclusion of members of disadvantaged communities on the bodies that set research priorities and apportion funding increases the likelihood of promoting a range of topics that will fairly represent the interests of those communities. Justice requires that we subject each level of research decision making to ethical review. By expanding the perspectives involved in making fundamental decisions about research programs, we are more likely to achieve the ethical ideal of a just research program.

Conclusion

A feminist lens guides us to take a broader perspective on the demands of ethics governing research with humans than the drafters of the *Belmont Report* adopted. Where the report accepted as given the background conditions under which research is pursued, a feminist lens focuses attention on the context of that research. It makes clear the importance of asking questions about justice at all stages of the research process. Only when the power to set the research agenda, determine process, decide on the selection of subjects, and control the emerging data is redistributed can we expect to have a truly just research process.

The commissioners who wrote the *Belmont Report* did constitute a group representing gender, racial, and disciplinary diversity. I credit this richness of experience and perspective with the degree of awareness present in the final report regarding many of the feminist concerns I have articulated. But lenses are more than uncritical windows on the world. Ideally, they are informed not only by personal experience but also by analysis of the ways in which things work. What is missing from the report is a distinctly feminist lens that focuses attention on the effects of relevant practices on oppressed social groups. Hence, the report repeatedly veers toward this level of analysis only to move back to the more established individualistic approach. This experience suggests that it is not sufficient, then, just to include women or minorities in the various decision-making bodies in ethics and science. It is necessary also to make certain that at least some participants have access to an explicitly feminist lens for their work.

The *Belmont Report* went a significant distance toward promoting more ethical research practices. It is now time to extend its vision and challenge the concentration of decision-making power in the hands of a privileged elite. Only when research programs are made accountable to all segments of the population can we be confident that subjects and citizens are treated ethically.

Notes

This chapter was initially prepared as a paper presented at the Rockefeller Foundation Bellagio International Study and Conference Center, Italy, March 2, 1999.

1. The feminist concerns I identify derive from my particular understanding of feminist ethics. See Susan Sherwin, *No Longer Patient: Feminist Ethics and Health Care* (Philadelphia, PA: Temple University Press, 1992). Feminists, like bioethicists, hold many diverse views, and I make no attempt to represent them all but simply identify those concerns that are particularly salient from my perspective.

2. I am particularly grateful to Françoise Baylis and Jocelyn Downie, who were especially diligent in helping to draft the Network's numerous interventions and comments on this subject.

3. We were mobilized to address this question in the hope of having an impact on a concurrent national effort to revise existing Canadian research ethics guidelines. The three major research funding agencies of Canada (the Social Sciences and Humanities Research Council, the Medical Research Council, and the Natural Sciences and Engineering Research Council) established a committee called the Tri-Council Working Group in 1994. This committee was asked to develop a common set of research guidelines to govern all publicly funded research involving humans in Canada. The Network communicated with the Tri-Council Working Group at various stages of their work in the hope that the emerging guidelines would meet the demands of our conception of feminist ethics. For a detailed discussion of the Network's efforts to ensure that the insights of feminist ethics were reflected in the Canadian research ethics guidelines, see Françoise Baylis, Jocelyn Downie, and Susan Sherwin, "Reframing Research Involving Humans," in *The Politics of Women's Health: Exploring Agency and Autonomy*, The Feminist Health Care Ethics Research Network, Susan Sherwin, Coordinator, 234–59 (Philadelphia, PA: Temple University Press, 1998), and Françoise Baylis, Jocelyn Downie, and Susan Sherwin, "Women and Health Research: From Theory, to Practice, to Policy," in *Embodying Bioethics: Recent Feminist Advances*, ed. Anne Donchin and Laura M. Purdy, 253–68 (Lanham, MD: Rowman and Littlefield, 1998).

4. B. Hammes and C. J. Laitman, "Diethylstilbestrol (DES) Update: Recommendations for the Identification and Management of DES-Exposed Individuals," *Journal of Midwifery and Women's Health* 48 (2003): 19–29.

5. J. Botting, "The History of Thalidomide," *Drug News and Perspective* 15 (2002): 604–11.

6. Writing Group for the Women's Health Initiative Investigators, "Risks and Benefits of Estrogen Plus Progestin in Healthy Postmenopausal Women," *The Journal of the American Medical Association* 288 (2002): 321–33.

7. Susan M. Wolf, "Erasing Difference: Race, Ethnicity, and Gender in Bioethics," in *Embodying Bioethics: Recent Feminist Advances*, ed. Anne Donchin and Laura M. Purdy, 65–81 (Lanham, MD: Rowman and Littlefield, 1998).

8. Ibid., 65.

9. There are other problems associated with oppression's impact on subjects' participation in research. See Susan Sherwin, "A Relational Approach to Autonomy in Health Care," in *The Politics of Women's Health: Exploring Agency and Autonomy* The Feminist Health Care Ethics Research Network, Susan Sherwin, Coordinator, 19–47 (Philadelphia, PA: Temple University Press, 1998). I argue there that autonomy must be conceived relationally to take account of the role of oppression in structuring the conditions under which oppressed people are asked to consent or refuse participation in research and therapy. The guidelines for informed consent in the *Belmont Report* do not allow for this complication.

10. The data to tell us how frequent this practice is are simply not available. What is clear is that it has occurred in several important areas of study (notably heart disease and AIDS, and it is not uncommon in other areas). See Anna C. Mastroianni, Ruth Faden, and Daniel Federman, eds., *Women and Health Research: Ethical and Legal Issues of Including Women in Clinical Studies*, 2 vols. (Washington, DC: National Academy Press, 1994).

11. Rebecca Dresser, "Wanted: Single, White Male for Medical Research," *Hastings Center Report* 22 (1992): 24–29.

12. Veronica Merton, "Review Essay: Women and Health Research," *Journal of Law, Medicine, and Ethics* 22, no. 3 (1994): 272–79.

13. The *Belmont Report* says at least that public money should not be used to support "development of therapeutic devices and procedures" that will be available only to those who can afford them.

14. Peter Singer, "Research into Aging: Should It Be Guided by the Interests of Present Individuals, Future Individuals, or the Species?" in *Life Span Extension: Consequences and Open Questions*, ed. Frédéric C. Ludwig (New York: Springer Publishing Company, 1991).

15. Barbara Katz Rothman, *The Tentative Pregnancy: Prenatal Diagnosis and the Future of Pregnancy* (New York: Viking Press, 1986).

16. See Barbara Katz Rothman, *Genetic Maps and Human Imagination* (New York: W. W. Norton and Company, 1998).

17. Public funds often are provided in partnership with private industry. Certainly, the training of researchers is largely subsidized by public money. And, in health care, the patients on whom therapies are tested are generally participating in a health care system with significant public support, even if the control resides in the hands of private corporations.

18. It would also explain how oppression interferes with autonomy such that informed consent per se is insufficient to protect autonomy. See note 9.

19. See Marilyn Frye, *The Politics of Reality: Essays in Feminist Theory* (Free-

dom, CA: Crossing Press, 1983); Linda Alcoff and Elizabeth Potter, eds., *Feminist Epistemologies* (New York: Routledge, 1993); Richmond Campbell, *Illusions of Paradox: A Feminist Epistemology Naturalized* (Lanham, MD: Rowman and Littlefield, 1998); and Sandra Harding, *Whose Science? Whose Knowledge? Thinking from Women's Lives* (Ithaca, NY: Cornell University Press, 1991).

20. The call for diversity does not mean that every imaginable perspective be represented in decision-making bodies, for that would be impractical. In addition, it allows specific exclusion of those who are committed to discriminatory social programs, as they do not support the underlying ethical principles at issue. Moreover, other positions will be well articulated in other forums or will not be relevant to the specific questions under review. But determining which voices must be present at each stage of decision making will always be a tricky ethical task. For example, occupational status will usually not be relevant, though it will matter greatly for certain sorts of occupational health and safety inquiries.

12 Protecting Communities in Research

From a New Principle to Rational Protections

Ezekiel J. Emanuel
Charles Weijer

Time is surely one of the greatest trials for any moral framework. Close to thirty years after the publication of the *Belmont Report*, it is time to look back in admiration at those who crafted the foundations upon which contemporary research ethics is firmly seated. The *Belmont Report*'s three ethical principles — respect for persons, beneficence, and justice — have served researchers, Institutional Review Boards (IRBs), and, most importantly, research subjects well. Novel diseases and technologies have tested the *Belmont* framework. The many moral challenges posed by human immunodeficiency virus (HIV) infection, thought at first to be novel, have largely been placed within the three principles.[1] Extraordinary advances in the technology to locate, sequence, and attempt to replace human genes have raised ethical quandaries that, after some reflection, are now recognized to fall for the most part within the established regulatory framework.[2]

But an anniversary is surely a time to look forward as well. We propose to do so regarding the protection of communities in research. Scientists are increasingly targeting whole communities in their search for the etiology of common diseases, including diabetes, asthma, and cancer. The prospect for benefit from such inquiry is considerable. It may lead to predictive tests and screening, better drug treatment, and perhaps novel cures. But such benefits will not come without risks to the very communities being studied. For instance, communities may be labeled as "diseased," resulting in discrimination. We will address two questions in this chapter: How ought we to situate issues related to the protection of communities in the moral framework of research ethics, and what sorts of additional protections, if any, ought to be afforded communities in research?[3]

How Ought We to Situate Issues Related to the Protection of Communities in the Moral Framework in Research Ethics?

Consider the following two research proposals.

Psychiatric illness in an Amish community. A geneticist proposes to study patterns of inheritance of bipolar affective disorder in a particular Amish community. The psychiatric disease is thought to be unusually common in this community, which may be due to the high degree of interrelatedness of its inhabitants. The geneticist will work in collaboration with a local psychiatrist who treats people from the community. Affected individuals will be invited to participate in the study by the treating psychiatrist. Data collection will include a medical history, family history, and blood sample for genetic analysis. Probands will be asked to invite other family members, whether they are affected or not, to participate in the study and submit a blood sample for analysis. Informed consent will be obtained from all individuals. The researchers hope the study will provide important information related to the location and structure of one or more genes for bipolar affective disorder.

Social predictors of tamoxifen use among women at risk for breast cancer. The use of tamoxifen to prevent breast cancer among women at risk remains controversial despite a number of recent reports of randomized controlled trials. A medical sociologist will examine the factors critical to women in their decision whether or not to use tamoxifen. A longtime member of Cancer Action Now (CAN), a regional breast cancer activist group, the sociologist will invite other members of CAN who do not have breast cancer themselves (e.g., children or siblings of persons who have the disease) to participate in the study. Participation will involve a ninety-minute semistructured interview. Questions will address risk of developing breast cancer, personal assessment of risk, and reasons for seeking out (or not) various cancer prevention strategies, including tamoxifen. Informed consent will be sought from all study participants. Researchers hope that the study will help identify whether women accurately assess their own risk of developing breast cancer and obstacles to the use of tamoxifen.

Both proposals target a particular community for study. Individual study participants are approached for participation because of their membership in an identified community. While both Amish and breast cancer activist groups are appropriately called "communities," important differences exist between them. The Amish are a cohesive community. They are situated in a discrete geographical location, and they share common religious beliefs, other values, history, and language. The community leader is a bishop elected by lot who speaks on their behalf on issues both religious and political. The breast cancer activist group is a less cohesive community. Women in CAN are united by a common goal, namely, improving the prevention of and treatment for breast cancer. Most members of the group either have the disease themselves or have an affected family member. However, members of CAN are geographically dispersed, espouse diverse religious beliefs, come from different cultural backgrounds, and do not necessarily share other values. While CAN possesses an elected board and president, these officials only represent the group on issues of policy related to the mission of the group.

While both studies pose risks to individual study participants, they also pose risks to the community as a whole. The first research proposal may profoundly affect the Amish community. The presence of researchers may be perceived as an unwanted intrusion into the community. Religious beliefs or other values may be challenged by information revealed to community members as part of the informed-consent process or the disposition of genetic samples, such as the creation of immortalized cell lines. The community as a whole may incur financial burden if researchers require accommodation in the community or if study participation requires that members take time away from work or other duties. Discovery and revelation of nonpaternity may lead to unwanted conflict within the community. Study results may perpetuate the false notion that psychiatric disease in general is more common in the Amish, and this in turn may lead to stigmatization and discrimination. Implications for the breast cancer activist community in the second study are, perhaps, less severe. This is due to the fact that a less cohesive community has fewer social structures that can be placed at risk. Nonetheless, if CAN is skeptical of drug-based cancer prevention strategies, the purpose of the study may conflict with the policy or values of the organization as a whole.

The ethical issues for the communities targeted in these two studies surely merit serious consideration. It is far from obvious, however, just how these quandaries fit within the moral framework articulated in the *Belmont Report*. Levine points out that the work of the National Commission in general has been criticized for overemphasizing individual rights.

> In each of its publications, [the National Commission] seems to embrace an atomistic view of the person. The person is seen as a highly individualistic bearer of duties and rights; among his or her rights, some of the most important are to be left alone, not to be harmed, and to be treated with fairness. Except, perhaps, in its report on research involving children, there is little or no reference to persons in relationship to others or as members of communities.[4]

The *Belmont Report* certainly seems guilty as charged. Indeed, the word *community* does not appear once in the document. This lack of attention to communities in research is further reflected in the federal Common Rule.[5] The regulations focus only on individual research subjects. The only groups considered are "vulnerable groups," such as the mentally ill and the educationally disadvantaged, and only insofar as they might be wrongfully included in or excluded from research participation.

It may be that the three *Belmont* principles have been interpreted too narrowly. Some have suggested that ethical concerns related to communities in research could be embraced under a broader understanding of one of the three existing principles. One suggestion is that community concerns might fit into an expanded notion of the principle respect for persons. Childress has suggested just this:

> Any serviceable account of biomedical ethics in a liberal society requires a central place for the principle of respect for autonomy. However, its demands are often unclear because of the complexity of personal actions and values, and because it is not the only source of moral guidance. In fact, the richest resolutions of debates in bioethics presuppose attention to the claims of other ethical principles as well as a fuller interpretation of selves in time and community.[6]

The enriched view of respect for persons has a number of advantages. First, it helps explain from where the values of individuals initially come. Without this understanding it becomes hard to explain how values ever develop. Second, it recognizes that many regard the relationship between the individual and community as an important part of their lives and that the community, and the values it embodies, is worthy of preservation.

However, an enriched version of respect for persons does not adequately account for the possibility of conflict between individual and communal choice. Imagine that protections for the Amish community in question require that consent be sought from the community before individuals are approached for study participation. The relationship between community and individual consent is asymmetric: both community and individual consent are required for a given person to participate, but community refusal overrides the preference of an individual community member. The demand of an individual community member to participate in a study regardless of the community's refusal is not captured by an all-encompassing notion of respect for persons. Gostin describes this conflict in terms of individual *versus* communal privacy rights:

> Could individuals in a study each give consent to disclosure of information, and yet the study would violate the right of the collective to privacy? The answer is yes if one believes that populations have a right to defend their reputation and dignity as much as individuals do. It is conceivable, even where the information revealed about each member of the group is non-consequential or is not personally identifiable, that the group can be harmed. . . . The right of populations to have some say in the collection and spread of data that they believe reflect badly on their identity is an important ethical principle.[7]

In other words, in a *Belmont* framework, such fundamental moral dilemmas are most naturally described and ultimately best resolved in terms of conflicts *between* principles.

Another possibility is that communal interests might be subsumed under an expanded version of the principle of beneficence. This view might begin by

pointing out that an individual's actions, desires, and goals are only comprehensible, and for that matter possible, within the context of a larger community. The desire for a fulfilling job presupposes a community with educational
institutions and employers. Improving one's health is possible only in the context of a community that aids in the distribution of basic needs, has appropriate health care institutions, and provides for the possibility of leisure time.
Thus, a vigorous and prosperous community as a whole is largely consistent
with the interests of individual members. On a deeper level, fulfillment of
basic individual interests presupposes various community institutions.

As mentioned earlier, however, this retooled version of the principle of
beneficence suffers from an inability to describe conflict between individual
and communal interests. Imagine that in the study involving the activist
group, one or more group members wish to participate because they want to
know whether their perceived risk of breast cancer corresponds to the actual
risk. Obtaining this information is straightforwardly in their interest: learning
that one is at lower risk than believed may reduce anxiety; learning that one is
at higher risk than perceived may allow one to seek out preventive treatments.
Participation in the study may not, however, be in the interest of CAN as a
community. If the group has taken a skeptical stance with regard to tamoxifen
as a drug to prevent breast cancer, the study's objectives may run afoul of the
group's policies and goals. It seems difficult to describe this conflict if individual and communal interests are subsumed under a single principle.

Moreno suggests a third possibility.[8] He believes that issues related to the
protection of communities in research are properly subsumed by the principle of justice. The history of research ethics is replete with examples of the
abuse of vulnerable groups. Moreno contends that communities in research
may be better thought of as vulnerable groups. For instance, in the Tuskegee
syphilis study, approximately four hundred poor, undereducated African
American men were deprived of treatment for syphilis so that researchers
could determine the course of untreated disease. In Moreno's analysis, this
group is both a community and a vulnerable group, and by implication communities are adequately protected by the principle of justice. Stemming from
the principle of justice, the Common Rule requires the IRB ensure that the

> selection of subjects is equitable. In making this assessment the IRB
> should take into account the purposes of the research and the setting in
> which the research will be conducted and should be particularly cog-
> nizant of the special problems of research involving vulnerable popula-
> tions, such as children, prisoners, pregnant women, mentally disabled
> persons, or economically or educationally disadvantaged persons.[9]

It is clear from this regulation that the term *vulnerable* encompasses a heterogeneous set of groups who are incapable of providing consent, not in a
position to provide free consent, or otherwise unduly susceptible to harm.[10]

But this suggestion too proves to be problematic. Although it may be
difficult to distinguish between vulnerable groups and communities at times,

not all communities are vulnerable groups, and not all vulnerable groups are communities. The protections to which Moreno refers have been shaped largely in reaction to unethical research involving vulnerable groups, such as children, the elderly, and racial minorities. Vulnerable groups are socially, economically, or otherwise disadvantaged and therefore are susceptible to exploitation or harm. Regulations protecting the vulnerable include added consent requirements and limits on the amount of nontherapeutic research to which they may be exposed.[11] Conversely, the driving force behind protections for communities is not vulnerability, but the fact that communities have values and interests worthy of protection. This implies a fundamentally different sort of relationship between community and researcher than between vulnerable group and researcher. The latter relationship is custodial in nature, in which the researcher assumes the responsibility to protect the vulnerable. In the case of a community, however, respect for the community calls not for a custodial relationship, but rather a partnership between community and researcher.

The Amish and children make the distinction between communities and vulnerable groups clear. The Amish are not a socially vulnerable group. While insular and voluntarily isolated from the larger society, they are not economically or otherwise disadvantaged. They are a community with values and interests deserving of respect and protection. Children are, by virtue of their inability to protect themselves in informed consent negotiations, a vulnerable group in need of custodial protections. While children are a vulnerable group, they are in no way a community.

None of the principles found in the *Belmont Report* seems adequate to capture the serious ethical issues raised by the inclusion of whole communities in research. We will not consider in detail the possibility of combinations of expanded principles. It may be that combinations of these principles might accommodate conflicts between individual values and communal interests and vice versa. The problem of conflicts between individual and communal values, or between individual and communal interests, is not, however, transparently solved by this approach. It therefore seems that we require a fourth principle for contemporary research ethics, a principle of respect for communities.

We are not the first to suggest the need for an additional *Belmont* principle. In 1982, Levine called for a principle of respect for cultures:

> Perhaps we should not attempt to develop universally valid standards for informed consent. Perhaps instead we should recognize the validity of certain forms of cultural relativism and have each culture decide how it should show respect for its own persons. If the guiding principle must have a name, I suggest that it might be "respect for cultures."[12]

More recently, McCarthy made a similar call for a principle of respect for families and communities, one that seems closer to the sort of cases under discussion:

> The principles have achieved notoriety in countries that stress individual liberty and place less emphasis on the rights and dignity of the com-

munity. However, many third world countries have much to teach developed nations about community values. For this reason I believe that the principles of justice, beneficence, and respect for persons should be supplemented by a fourth principle, dealing with respect for the family and the community of the research subjects.[13]

Unfortunately, neither proposal is accompanied by detailed argumentation as to why such a principle ought to be adopted or details as to how it would be implemented.

The principle of respect for communities is reasonably interpreted as conferring upon the researcher an obligation to respect the values and interests of the community in research and, wherever possible, to protect the community from harms. Indeed, a fourth principle has considerable pragmatic appeal. Cases involving conflict between individual and communal choices, or between individual and communal interests, can be described and analyzed in terms of conflicts between the principles of respect for persons and respect for communities, and beneficence and respect for communities, respectively. Also, issues pertaining to vulnerable groups and communities may now be separated under principles of justice and respect for communities. As we have said, each implies a differing relationship between researcher and social group involved: researchers ought to have a custodial relationship with vulnerable groups, and they ought to have a partnership with communities. Insofar as an individual community happens to be vulnerable, tension exists between these two models, and it can be described as a conflict between demands of the principles of justice and of respect for communities.

These are pragmatic reasons for adding a principle of respect for communities. But there are also substantive reasons. The prime effect of a principle of respect for communities is to acknowledge that the community is more than the sum of individual values and interests; the community itself has values and interests. For that reason, it is appropriate that it be accorded respect separate from that owed individual community members. In short, the community ought to be accorded moral status. The reasons for this are multiple. First, people do not view themselves atomistically; they view themselves as members of one or more communities that constitute their values and self-understandings. Second, various communities are already given the authority to make binding decisions on behalf of individual members. Thus, the state may levy taxes, the municipality may decide which school a child may attend, an Amish bishop may decide when a crop is to be planted, and a cancer activist group president may speak on behalf of her membership to the press. Third, the primacy of the individual versus the community varies from one community to the next in the world. It may be that the individual is prime in Western liberal states, but in certain communities even within these states, such as the Amish, this is not the case. In these communities, the community itself is legitimately viewed as prime.

As with any novel principle, one must ask whether a principle of respect for communities may be used to perpetuate immoral practices. For instance, confronted with our proposal for a principle of respect for communities, Faden asked, "What about women?"[14] That is, might a principle of respect for communities be used as a justification to perpetuate the oppression of groups such as women within certain communities? A community with a male-dominated leadership may silence the voices of women. For instance, community leaders may be reluctant to permit research into the prevalence and causes of spousal abuse because the results may reflect poorly on the community as a whole, or at least the men within it. This is an important problem and one that a legitimate moral framework must be able to solve. It is crucial to understand that we have not proposed to replace the three existing *Belmont* principles with a new principle; rather we wish to add a fourth principle to the existing three. Thus, the force of all four principles can be brought to bear on this and similar problems. One formulation of justice requires that the burdens and benefits of research participation be distributed equitably; another aspect of justice calls for the elimination of domination.[15] A community that seeks to perpetuate oppression is hence legitimately criticized on grounds of justice, and community safeguards hijacked to perpetuate such oppression have no moral force.

What Sorts of Additional Protections Ought to Be Afforded Communities in Research?

In the preface to *A Theory of Justice*, Rawls states:

> During much of modern moral philosophy the predominant systematic theory has been some form of utilitarianism. One reason for this is that it has been espoused by a long line of brilliant writers who have built up a body of thought truly impressive in its scope and refinement. . . . [Conversely] those who criticized them often did so on a much narrower front. They pointed out the obscurities of the principle of utility and noted the apparent incongruities between many of its applications and our moral sentiments. But they failed, I believe, to construct a workable and systematic moral conception to oppose it.[16]

We confront much the same situation regarding the three *Belmont* principles. Unless we can provide a "workable and systematic" conception of the principle for respect for communities, it is likely to remain an amorphous and impotent notion, inciting conflict while lacking a practical vision that can be implemented. Without a systematic vision of what respect for communities means in practice that is sensitive to the different types of communities, the three principles alone will remain the foundation of research ethics.

To offer this "workable and systematic" conception we need to address some fundamental questions: What is a community? How should different communities be distinguished? How can appropriate protections be mapped onto the different types of communities? What constitutes community consent or consultation? A five-step approach will provide answers to these questions and a practical vision of this new principle. The five-step approach entails (1) *identifying* community characteristics relevant to the biomedical research setting, (2) *delineating* a typology of different types of communities using these characteristics, (3) *summarizing* the range of possible community protections, (4) *connecting* particular protections to specific community characteristics necessary for its implementation, and (5) *synthesizing* community characteristics and possible protections to define protections appropriate for each type of community.

Step 1: Identifying relevant community characteristics. "Community" is a polyvalent term with a wide range of meanings. It can be used to refer to a wide variety of human associations, from tribes to unions and from municipalities to religious adherents. Because communities differ, protections for them will also need to vary. A single set of regulations to fit all types of communities is doomed to failure. What are needed are morally relevant criteria that distinguish communities. Ten characteristics relevant to communities in biomedical research can be identified, including a health-related common culture, legitimate political authority, and communication network. This list of community characteristics is neither an exhaustive list nor a specification of necessary and sufficient conditions for what constitutes a community. Rather, these characteristics serve as a practical basis for the construction of a typology of communities restricted to health-related areas.

Step 2: Delineating a typology of differing types of communities. Using these ten characteristics, a typology distinguishing seven different types of communities can be delineated: aboriginal, geographic or political, religious, disease, ethnic or racial, occupational, and virtual communities. Not all these communities are composed of all ten characteristics. Communities may be arrayed along a spectrum of cohesiveness, from those that possess all the characteristics to those that possess only a few. At one end of the spectrum, a cohesive aboriginal community often — or nearly always — has all of the characteristics listed. Conversely, a less-cohesive occupational community embodies only two of the characteristics: common culture and a communications network. This list of seven types of communities is neither exhaustive nor comprehensive; there may well be other types of communities. Rather, this list is meant to cover key communities that may be involved in biomedical research. An effective strategy of identifying protections for communities in research will therefore need to take into account the spectrum of communities and the morally relevant differences among them.

Step 3: Summarizing the range of possible community protections. Having delineated a typology of communities, the next step is to identify a comprehensive set of potential protections. A survey of seventeen national and international documents detailing protections for particular communities revealed five broad themes encompassing twenty-three specific protections.[17] They extend from the genesis of the research project to publication of the results. The

TABLE 12.1 Morally relevant characteristics of communities in
biomedical research

Community Characteristic	Definition
Common culture and traditions, cannon of knowledge, and shared history	The community shares cultural elements, including traditions, a body of knowledge, language, and history that all or many members practice, know, speak, or identify with.
Comprehensiveness of culture	Culture pervades many aspects of the life of community members, including broad notions of how to live one's life well or properly; it is not restricted to just one domain of activity, such as work.
Health-related common culture	The community shares ideas regarding the nature and treatment of disease, as well as health priorities.
Legitimate political authority	The community is represented by one or more officials who are accorded, through a legitimate political process, the authority to speak for, and make binding decisions on behalf of, the community across a range of domains.
Representative group or individuals	The community has a group (or an individual) that speaks on their behalf, reports information back, and is held accountable by it.
Mechanism for priority setting in health care	The community has a system, usually a particular organization within it, that makes decisions (or develops plans for approval) regarding the allocation of health care resources.
Geographic localization	The community is located in one geographic area, such as a town, settlement, or county.
Common economy or shared resources	The community has a system for the exchange and distribution of goods; the prosperity of community members depends in part on that of the community as a whole.
Communication network	There exists some mechanism for information to be shared among community members, be it by a meeting in a town hall, a newspaper, a newsletter, or via the Internet.
Self-identification as a community	Individuals self-identify as being members of the community with its cultural, social, and ethical norms informing their beliefs and actions.

TABLE 12.2. The characteristics of various types of communities
in biomedical research

Community Characteristic	Type of Community						
	A	G/P	R	D	E/R	O	V
Common culture and traditions, cannon of knowledge, and shared history	++	+	++	+/-	+	++	+
Comprehensiveness of culture	++	+/-	++	-	+	+/-	-
Health-related common culture	++	+	++	++	+	+/-	-
Legitimate political authority	++	++	+/-	-	-	+/-	-
Representative group/individuals	++	++	++	+	+	+/-	+/-
Mechanism for priority setting in health care	+	+	+/-	+	+/-	+/-	-
Geographic localization	+	++	+/-	+/-	+/-	-	-
Common economy/ shared resources	++	++	+/-	+/-	+/-	-	-
Communication network	++	+	+	+/-	+/-	+	++
Self-identification as community	++	++	++	+/-	+	+/-	+

Legend

A   Aboriginal
G/P   Geographic/Political
R   Religious
D   Disease
E/R   Ethnic/Racial
O   Occupational
V   Virtual
++   Community *nearly always or always* possesses the characteristic
+   Community *often* possesses the characteristic
+/-   Community *occasionally or rarely* possesses the characteristic
-      Community *very rarely or never* possesses the characteristic

themes and individual guideline requirements for the protection of communities in research are as follows:

- Consultation in protocol development: respect for the community's culture; community input on protocol development; research is useful to the community; respect for the community's knowledge and experience
- Information disclosure and informed consent: nontechnical and appropriate disclosure; face-to-face meetings with the community; adequate

time for community review; community consent; community consent required for protocol changes

- Involvement in research conduct: transfer of skills and research expertise to members of the community; employment for members of the community; reimbursement to the community for research costs; informing the community about research progress
- Access to data and samples: community consent for further use of samples; storage of data negotiated
- Dissemination and publication of results: involvement of the community in manuscript preparation; draft report transmitted to the community for comment; acknowledgment of the community; consent to identify the community; report compliance with guidelines; final report provided to the community; community consent for research media interview

While this is not an exhaustive list of protections for communities in biomedical research, it is a reasonably comprehensive listing of protections articulated to date. It is necessary to determine which of these protections is appropriate for particular types of communities.

Step 4: Connecting particular protections of specific community characteristics necessary for its implementation. More than forty years ago the legal theorist Lon Fuller stated in a famous set of lectures that the purpose of the law is "the governance of human conduct by means of rules."[18] He argued that the law must have at least eight necessary characteristics to realize this end, for example, laws cannot be retroactive but must be prospective, and laws must be promulgated not secret. Adopting an approach similar to Fuller's, it is possible to identify particular characteristics of communities that are necessary for the implementation of specific protections. In this way, each of the identified community characteristics (table 12.1) is linked to one or more of the protections (table 12.3).

An example from each of the five themes of community protections illustrates the deductive reasoning used. If the community is to have input on the protocol, the community must have representatives who can provide this input on behalf of the community. Similarly, if community consent is to be sought before individuals are approached for study participation, the community must have a legitimate political authority that is empowered to speak authoritatively for and make binding decisions on behalf of the community; more than mere representation is required. In addition, if the community is to be reimbursed for research costs, the community must have a common economy or shared resources. Furthermore, community consent for further use of samples requires not only that the community have a legitimate political authority but also that they have a health-related common culture. Unless there are shared ideas about health, the disposition of samples is likely to be of little or no relevance to the community as a whole, although partic-

TABLE 12.3 Community characteristics necessary for the implementation of particular protections

Proposed Protections	Community Characteristics							
	CC	PA	RG	PS	GL	CE	CN	SI
Consultation in protocol development								
Respect for culture	×							
Input on protocol			×					
Research useful				×				
Respect for knowledge and experience	×							
Process of providing information and obtaining informed consent								
Nontechnical and appropriate disclosure			×				×	
Face-to-face meetings			×		×			
Adequate time for review			×					
Consent		×						
Consent required for protocol changes		×						
May withdraw consent		×						
Involvement in research conduct								
Transfer of skills and expertise						×		
Employment						×		
Reimbursement for research costs						×		
Informed about research progress			×				×	
Access to data and samples								
Consent for further use of samples	×	×						
Storage of data negotiated			×					
Dissemination and publication								
Involvement in manuscript preparation			×					
Draft report for comment							×	
Acknowledgment								×
Consent to identify		×						
Report compliance with guidelines								
Final report			×				×	
Consent for researcher media interview		×						

Legend

CC   Health-related common culture
PA   Legitimate political authority
RG   Representative group or individuals
PS   Mechanism for priority setting in health care
GL   Geographic localization
CE   Common economy or shared resources
CN   Communication network
SI   Self-identification as a community

ular individuals within it may have strong feelings. Finally, if a meaningful draft report for comment is to be provided to the community, it must possess a communication network so that the report can be distributed to community members.

This delineation of community characteristics and their attendant protections determines the potential protections for the different types of community. Thus, the characteristics a community possesses circumscribe a boundary of potential protections.

Step 5: Synthesizing community characteristics and possible protections to define protections appropriate for each type of community. Thus far, different types of communities have been distinguished using characteristics relevant to health research. Specific protections were then linked with particular community characteristics. Through community characteristics it is now possible to link specific protections with each distinct type of community. For instance, communities possessing a common economy or shared resources can have transfer of skills or expertise to the community, employment for community members, and reimbursement for research costs incurred by the community. Similarly, communities that have both a representative group and a communication network can have input on the protocol, nontechnical and appropriate disclosure of information related to the study, adequate time for review, information about study progress, negotiations about storage of data or biological samples, involvement in manuscript preparation, and a final report of the research finding. Proceeding in this way, three general regimes of protection for communities in research can be delineated: (1) community consent, (2) community consultation alone, and (3) no added protections.

Community consent is only possible if the community has a legitimate political authority. Community consent can only occur legitimately if the community has conferred upon some person or group — be it a legislative assembly, mayor, tribal council, or bishop — the authority to make binding decisions on behalf of its members. Without such an authority, no person or group has the political or moral authority to consent legitimately of behalf of the collective. For instance, the breast cancer activists have no legitimate political authority, and hence suggesting that community consent be sought from them — or any other community lacking a legitimate political authority — is neither morally nor pragmatically justifiable.

But this does not undermine the importance of respect for communities or the possibility that community consent is appropriate for some communities. Different types of protections are appropriate and feasible only when communities have certain characteristics, and different communities will perforce have different types of protections.

More than the mere feasibility of protections for particular communities is at stake, however. Current guidelines extend the full list of protections to aboriginal communities (table 12.3). Many geographic and political commu-

TABLE 12.4 Checklists of proposed protections for different types of
communities in biomedical research

	Community Consent and Consultation	Community Consultation Alone	No Added Protections
Types of communities included {	Aboriginal, geographic/ political	Religious, disease, ethnic/racial	Occupational, virtual

Proposed Protections

Consultation in protocol development

Respect for culture	×	×	
Input on protocol	×	×	
Research useful	×	×	
Respect for knowledge and experience	×	×	

Process of providing information and obtaining informed consent

Nontechnical and appropriate disclosure	×	×	
Face-to-face meetings	×	×	
Adequate time for review	×	×	
Consent	×		
Consent required for protocol changes	×		
May withdraw consent	×		

Involvement in research conduct

Transfer of skills and expertise	×		
Employment	×		
Reimbursement for research costs	×		
Informed about research progress	×	×	

Access to data and samples

Consent for further use of samples	×		
Storage of data negotiated	×	×	

Dissemination and publication

Involvement in manuscript preparation	×	×	
Draft report for comment	×	×	
Acknowledgment	×		
Consent to identify	×		
Report compliance with guidelines	×	×	
Final report	×		
Consent for researcher media interview	×		

nities, including the Amish community in our first example, are almost identical to aboriginal communities in terms of morally relevant criteria. It would be unjust not to confer similarly comprehensive protections to these communities. Other communities, such as religious, disease, and ethnic or racial characteristics, including CAN in our second example, may not be identical to aboriginal communities but share with them some morally relevant characteristics, such as community representatives, health care common culture, and a communication network. It follows that added protections ought to be conferred on these differing communities insofar as they share relevant characteristics with aboriginal communities. The existence of well-accepted guidelines for the conduct of research in one on these communities, the HIV community, also supports this claim.[19] In general, only aboriginal and geographic or political communities — and the Amish community referred to earlier — possess an authority that can make binding decisions on behalf of its members, such as community consent. Because other types of community, including religious, disease, ethnic or racial, occupational, and virtual communities, lack legitimate political authority, it would be inappropriate to require the consent of the community before individual research subjects are approached for consent. Not surprisingly, those communities that possess legitimate political authorities are among the most cohesive communities, possessing all or most of the characteristics relevant to the implementation of guidelines requirements (table 12.2). Thus, protections for aboriginal and geographic or political communities in research include the full list of guideline requirements; in other words, both community consent and consultation are required (table 12.4).

So then, for the study involving the Amish community, the full range of community protections is appropriate. The community, for instance, must be involved in study design, community consent must be sought, the community must be involved in the conduct of the study, community consent must be sought for use of the samples after the study, and the community must be involved in the preparation of the final report. The researcher in this instance shows respect for the community by seeking its consent and involvement.

Even though community consent is not possible for the breast cancer activist community, other protections that may be characterized as "community consultation" are appropriate. Community consultation encompasses the involvement of community representatives to a limited degree in study planning, informing the community as a whole of the study at its start and as progress is made, consulting with community representatives regarding the disposition of data, and providing them with a draft report on which to comment. Communities such as the breast cancer activists — religious, disease, and ethnic or racial communities — may be relatively cohesive and share many, though not all, of the characteristics required for the implementation of guideline requirements. Reflecting the intermediate degree of cohesiveness, the list of potential protections listed is roughly half that for aboriginal and geographical or political communities (table 12.4). Occupational and virtual communities are the least cohesive of the communities in the typology (table 12.2). Generally, they possess few of the morally relevant community characteristics, and, accordingly, no added protections are required for research involving them. However, there may be exceptions.

Importantly, the communities we have delineated are ideal types and are not meant to reflect any actual communities.[20] Within actual communities of each type there will be diversity. Particular communities may consequently constitute exceptions to the general scheme delineated. For instance, farm workers or coal miners with strong union representation, geographic localization, union-based health insurance, and other social security programs may be sufficiently cohesive — that is, possess the ten characteristics — as to legitimize the additional protections of community consent and consultation. Similarly, while the Amish may be a religious community, they are sufficiently cohesive and have legitimate community leaders empowered to make health care decisions that, as a community, they require the protections of community consent.

In addition, it is important to recognize that human associations are not static but dynamic. Bonds within a group may strengthen over time, and novel social structures may emerge as a new community develops, necessitating reconsideration of the level of protections. For example, it is not meaningful to speak of many disease groups, such as asthmatics or bald men, as communities; they share few if any of the characteristics that bind one to another in a community. However, at some time in the future they may come together for support and to advocate for more research funding and a voice in setting a research agenda. Local groups may coalesce into a national advocacy organization with elected regional representation, policies outlining health research priorities may be adopted, and a monthly newsletter and email list server may enable communication within the group. Over time, individuals may begin to coalesce into informal groups and later metamorphose into a community. In this evolution, such a community accrues these same morally relevant characteristics that confer the ability and obligation to enact protections when the community is targeted for research. Similarly, communities can disintegrate. A relatively cohesive community may experience disagreements, rupture, and disintegration over time, losing characteristics and thereby losing attendant protections.

Conclusion

For any written work to serve as a practical guide nearly thirty years after its publication is remarkable and signifies the enduring insights of the document. But the *Belmont Report* was written under the grip of an individualist vision. New challenges and greater appreciation of the ways in which individuals are embedded in communities have made its approach seem limited. We have argued that the addition of a fourth ethical principle — respect for communities — supplements the *Belmont* framework for research ethics, enhancing its applicability to current research studies and challenges. Simultaneously, this principle will founder unless the practical implications of the principle can be delineated in a systematic manner. The five-step analysis applies the principle to actual practice. The real key to the practical application of the principles is to recognize that not all groups commonly referred to as communities are the same. The typology is a pragmatic mechanism to distinguish different communities. Protections can then be mapped onto particular communities that possess requisite characteristics.

This rational strategy can guide researchers, communities, and IRBs in determining appropriate protections for particular communities. Inevitably, this rational procedure requires interpretation as it is implemented at the local level. Investigators and IRBs will have to decide whether a group constitutes a community, what type of community, and what protections are appropriate. Just as determining whether particular research is minimal risk or not requires interpretation and judgment, so too does determining whether added protections for a particular community are needed and what they ought to be.

Researchers approaching a community regarding a novel research project can adopt this method by determining the characteristics of the community and the ensuing scope of issues to be discussed. Similarly, a community that is approached frequently for study participation may use this framework (table 12.4) as a starting point for self-generated guidelines and a framework for their relations with researchers. Furthermore, an IRB reviewing a research protocol involving a community may compare proposed protections in a study involving a particular community with a complete list of possible protections to identify inappropriate protections or neglected areas of potential concern. All of these ways of implementing protections for community — by researchers, communities, and IRBs — can occur now and without additional regulations. After dialogue, regulators can see if consensus evolves toward the need for regulation. Should federal regulators decide to codify protections for communities in regulation, the approach this chapter describes should help in defining appropriate protections for communities in research.

Notes

1. Carol Levine, Nancy N. Dubler, and Robert J. Levine, "Building a New Consensus: Ethical Principles and Policies for Clinical Research on HIV/AIDS," *IRB: A Review of Human Subjects Research* 13, nos. 1–2 (1991): 1–17.

2. K. C. Glass, C. Weijer, R. Palmour et al., "Structuring the Review of Human Genetics Protocols: Gene Localization and Identification Studies," *IRB: A Review of Human Subjects Research* 18, no. 4 (1996): 1–9; K. C. Glass, C. Weijer, C. T. Lemmens et al., "Structuring the Review of Human Genetics Protocols, Part II: Diagnostic and Screening Studies," *IRB: A Review of Human Subjects Research* 19, nos. 3, 4 (1997): 1–11, 13; K. C. Glass, C. Weijer, D. Cournoyer et

al., "Structuring the Review of Human Genetic Protocols, Part III: Gene Therapy Studies," *IRB: A Review of Human Subjects Research* 21 (1999): 1–9.

3. Charles Weijer, "Protecting Communities in Research: Philosophical and Pragmatic Challenges," *Cambridge Quarterly of Healthcare Ethics* 8 (1999): 501–13; C. Weijer, G. Goldsand, and E. J. Emanuel, "Protecting Communities in Research: Current Guidelines and Limits of Extrapolation," *Nature Genetics* 23 (1999): 275–80; Charles Weijer and Ezekiel J. Emanuel, "Protecting Communities in Biomedical Research," *Science* 289 (2000): 1142–44.

4. Robert J. Levine, *Ethics and Regulation of Clinical Research*, 2nd ed. (New Haven, CT: Yale University Press, 1988), 13.

5. 45 C.F.R. 46.

6. James F. Childress and John C. Fletcher, "Respect for Autonomy," *Hastings Center Report* 24, no. 3 (1994): 34–35.

7. L. Gostin, "Ethical Principles for the Conduct of Human Subject Research: Population-Based Research and Ethics," *Journal of Law, Medicine & Health Care* 19 (1991): 191–201.

8. Personal communication with Jonathan Moreno, April 1999.

9. 45 C.F.R. 46.111 (a)(3).

10. Charles Weijer, "Research Involving the Vulnerable Sick," *Accountability in Research* 7 (1999): 21–36.

11. Charles Weijer, "Thinking Clearly about Research Risk: Implications of the Work of Benjamin Freedman," *IRB: A Review of Human Subjects Research* 21, no. 6 (1999): 1–5.

12. Robert J. Levine, "Validity of Consent Procedures in Technologically Developing Countries," in *Human Experimentation and Medical Ethics: Proceedings of the 15th CIOMS Round Table Conference*, ed. Z. Bankowski and N. Howard-Jones, 16–30 (Geneva: Council for International Organizations of Medical Sciences, 1982).

13. Charles McCarthy, "A North American Perspective," in *Ethics and Research on Human Subjects: International Guidelines*, ed. Z. Bankowski and R. J. Levine, 208–11 (Geneva: Council for International Organizations of Medical Sciences, 1993).

14. Personal communication with Ruth Faden, April 1999.

15. Charles Weijer, "Selecting Subjects for Participation in Clinical Research: One Sphere of Justice," *Journal of Medical Ethics* 25 (1999): 31–36.

16. John Rawls, *A Theory of Justice* (Cambridge, MA: Harvard University Press, 1971), vii–viii.

17. Weijer, Goldsand, and Emanuel, "Protecting Communities in Biomedical Research."

18. Lon L. Fuller, *The Morality of Law* (New Haven, CT: Yale University Press, 1964).

19. Levine, Dubler, and Levine, "Building a New Consensus."

20. Max Weber, *The Theory of Social and Economic Organizations*, ed. Talcott Parsons (New York: Free Press, 1947).

13 Ranking, Balancing, or Simultaneity

Resolving Conflicts among the *Belmont* Principles

Robert M. Veatch

After thirty years, it is fitting to explore how successful the *Belmont Report*[1] was in articulating the basic principles of human subjects research and to examine some problematic implications as well as questions left unanswered. If there was ever an example of twenty pages of philosophical ethics that have left an indelible mark on the public, it is surely *Belmont*. That it is still cited by philosophers, researchers, institutional review boards (IRBs), and courts as one of the most important, if not the definitive, codifications of the principles of research ethics testifies to its importance. At the same time there remain some outstanding questions that must be resolved if human subjects are to be adequately protected and human subjects research projects are to be established as ethical.

The *Belmont Report* is clearly within what has now come to be known as the principlist camp in bioethics.[2] It articulates a small number of very general principles for guiding actions and/or rules of conduct while acknowledging that there may not be consensus on underlying foundations,[3] on the more concrete norms of practice favored by the casuists,[4] or on the character traits deemed virtuous in those involved in human subjects research.[5] *Belmont* is testament to how far ethics can go without any explicit agreements on these other areas of ethics.

Belmont also reveals the potential for variation among principle-based theories. Most notably, the report identifies three principles (respect for persons, beneficence, and justice), rather than adopting any of the other lists of principles — which often produce a different count. *Belmont* thus appears to position itself as an alternative to the single-principle theories of ethics (whether utilitarian or libertarian). It can also be contrasted with those who favor two-principle (consequentialists, who believe there is a difference of more than direction between beneficence and nonmaleficence, as well as other forms of two-principle theories such as Engelhardt's affirmation of the principles of permission and utility) or four-principle approaches (Beauchamp and Childress or Gillon). It can also be contrasted with the five "moral appeals" of Baruch Brody[6] (not all of which appear to be principles), as well as the six prima facie duties of W. D. Ross[7] and the ten "moral rules" of Bernard Gert[8] (both of which have many of the characteristics of principles). It stands in contrast with my own seven principles.[9] Hence, there is room for variation among principlists, and *Belmont* is potentially in tension with each of these authors of alternative lists, at least on some marginal matters. This suggests several issues yet to be addressed if *Belmont* is to become the definitive set of principles for human subjects research. I first look at five issues I term "preliminary" in the sense that, although they identify problems with *Belmont*, they are problems that should be resolvable with clarification in terminology or emendations to the theory. Then I turn to what I consider to be the single most crucial problem with *Belmont*: its failure to make clear what should happen when the principles conflict among themselves, that is, when, in the assessment of a proposed protocol, one principle supports one conclusion about a protocol while another leads to a different conclusion. Do all the principles have to be satisfied simultaneously for a protocol to be acceptable, or must the principles be satisfied merely "on balance"? In particular, can enough anticipated benefit from a protocol (beneficence) justify compromises with the principles of respect for persons and justice?

I will refer to the view that all the principles must be satisfied in order for a protocol to be morally acceptable as the *simultaneity view*. All the principles are independently necessary conditions, and only when all are satisfied are the sufficient conditions for justification met. Conversely, I will refer to the notion that the principles must merely be satisfied on balance as the *balancing view*. There is a third possibility, which I will call the *ranking view*. It holds that principles can be rank-ordered (lexically ordered, to use Rawls's expression) and that the highest-ranking principle must be fully satisfied before the next in rank is considered. The morally superior course is the one that moves furthest down the list without violating any higher-ranked principles. I will argue for a combination of the *simultaneity* and *ranking* views and against the *balancing* position. I note that, in one sense, it seems obvious that the principle utility (the National Commission's "beneficence") must be satisfied. No amount of respect for autonomy and justice in a protocol would justify research that was expected to do more harm than good. Thus, whatever we say about simultaneity and the other principles, this formulation of utility must be satisfied. On the other hand, if the principle of utility is formulated, as it is in some utilitarian theories, to mean that, for a research design to be moral, it must be expected to maximize utility, then almost no protocol that concerns itself with either autonomy or justice would satisfy utility. Any investigator trying to maximize the benefit of his or her research would prefer not to give subjects the chance to refuse to participate and that investigator would also prefer not to be constrained in subject selection by the requirements of justice. Almost no protocol considered ethical by today's standards is truly a protocol that is as elegant as it could be in terms of being the best possible science. Almost none could be said to lead to maximum social benefit when compared to protocols that researchers could imagine if they were not constrained by respect for persons and justice. Hence, if utility is understood to mean that there must be positive utility, it is surely a necessary condition for research to be ethical. If it is understood to mean that the protocol must be expected to produce the maximum information with maximal elegance, then some compromise with the principle will almost certainly be necessary in most cases.

Preliminary Problems

Before turning to the critical problem of whether all the *Belmont* principles must be satisfied simultaneously for a protocol to be morally acceptable, five preliminary issues need to be clarified. Some of these have been recognized almost from the moment *Belmont* was issued. Others, while not receiving a great deal of attention, seem to suggest clear answers once they are made visible.

THE RELATION OF UTILITY TO BENEFICENCE AND NONMALEFICENCE

The first and most conspicuous issue raised by the *Belmont* principles is why the National Commission chose to use the term *beneficence* for one of the principles rather than including an additional principle of *nonmaleficence* or, alternatively, referring to the principle of *utility* as a way of covering both beneficence and nonmaleficence. Beneficence means literally "doing good," yet the report claims that two general rules are entailed: do not harm, and maximize possible benefits and minimize possible harms.

The use of *beneficence* in such a way that it encompasses both doing good and avoiding harm is certain to lead to confusion in IRBs and other moral assessments of research protocols. Some moral theories give special priority to not harming, considering it more weighty or imperative than doing good.[10] Others consider doing good and avoiding harm to be merely two dimensions of the same variable, one positive and the other negative. According to this second view, benefits and harms can be combined on the same scale. (Even then there is controversy over whether the two should be combined arithmetically — subtracting harms from benefits — or geometrically — calculating the ratio of benefits to harms. The two approaches can yield dramatically different results.)

The correction of *Belmont* seems obvious. If one believes that doing good and avoiding harm are merely two poles of a utility calculation, then the misleading term *beneficence* should be replaced with a term that avoids implying that only the positive dimension is being considered. The principle of utility is the obvious choice; then all that would be needed is an explication of how benefits and harms are to be combined into a single measure of net consequences. If, on the other hand, one believes that there are meaningful conceptual or moral differences between doing good and avoiding harm — if, for instance, one believes like Ross that avoiding harm is a more stringent duty —

then one needs to add an additional principle to the *Belmont* troika so that both beneficence and nonmaleficence are included. Beauchamp and Childress handle this problem in a similar way.

THE WELFARE OF CHILDREN

Second, almost everyone seems troubled by *Belmont*'s handling of children and other incompetent persons. The report places the duty to protect the interests of those with diminished autonomy under the principle of respect for persons. According to the *Belmont Report*, respect for persons incorporates at least two basic ethical convictions: that individuals should be treated as autonomous agents and that persons with diminished autonomy are entitled to protection.

While I concur that there is more to respect for persons than merely respecting the autonomy of substantially autonomous agents (a point I will take up next), it is not clear why the duty to protect the substantially nonautonomous falls under this principle as well. *Belmont* claims that some persons are in need of protection from activities that may harm them: "the extent of protection afforded should depend upon the risk of harm and the likelihood of benefit." While that claim is not controversial, the suggestion that this is a specification of the principle of respect for persons is controversial. It seems much more plausible to make protection of the interests of the incompetent an entailment of the principles beneficence and nonmaleficence (or utility, if the two are combined).

AUTONOMY VERSUS RESPECT FOR PERSONS

Third, although the *Belmont* authors may have been mistaken in subsuming protection of the interests of incompetents under respect for persons, they are probably correct in their claim that there is more to this principle than merely respecting the autonomy of persons insofar as they are autonomous. What is missing from *Belmont* is an explication of what else is entailed in this principle besides respecting autonomy.

I am convinced that there are at least three other implications. These can be summarized under the headings of fidelity, veracity, and avoidance of killing. In *A Theory of Medical Ethics* I claimed that these three plus autonomy constituted four separate principles that took morality beyond "consequence-maximizing." Independent of maximizing net utility, morality, according to the position developed in that work, also involves duties of fidelity to commitments, telling the truth, and avoiding killing (at least avoiding killing humans). The claim placed me in the deontological camp of W. D. Ross and Immanuel Kant.

I am now prepared to change the language somewhat and refer to these four (fidelity, veracity, avoidance of killing, and respect for autonomy) as aspects of an overarching principle of respect for persons. I don't think it matters much whether one says that there is one principle with four separate manifestations or that there are four principles all conveying aspects of respect for persons. What is critical is that there is more to morality than merely maximizing net utility and respecting autonomy. This means that the *Belmont* authors did not go far enough. They did not provide an explicit basis for the duty of investigators to keep promises, tell the truth, and avoid killing.

One can imagine potential subjects who are not autonomous agents and who will not plausibly be harmed by their incorporation into a research project. If we are limited to *Belmont*, with its restricted explication of the respect for persons principle, I see no basis in the report for investigators to speak truthfully to such subjects or keep promises made to them. I don't think there is an adequate basis for a duty to keep medical information confidential unless it is grounded in promises made. All of these duties seem to be clearly entailed in the meaning of respect for persons. Only by expanding respect for persons so that it is a "super-principle" entailing these other moral considerations would there be an adequate basis for grounding the full range of duties widely recognized as incumbent on investigators, IRB members, sponsors, and others involved in human subjects research.

JUSTICE IN RESEARCH DESIGN AND EXECUTION

A fourth emendation to the *Belmont Report* centers on problems related to the authors' understanding of the principle of justice. *Belmont* was the first public medical ethics document to explicitly recognize that justice is morally relevant in human subjects research. Nevertheless, what was an advance in its time is now, thirty years later, terribly inadequate. After providing a rich introductory paragraph on the principle of justice, *Belmont* offers a very restricted explication of the implications for human subjects research. It was pioneering in stressing the implications for selection of research subjects, which, in fact, is the sole topic related to justice in the section of the report dealing with the implications of the three principles. It forcefully challenges investigators to avoid discriminatory selection of subjects from the poor, the institutionalized, and racial or ethnic minorities. In the initial discussion of the principle of justice there is also a brief mention that research supported with public funds should be made available equitably.

What is missing is any implication of the principle of justice for the design and execution of research. This has long been a concern of mine.[11] The principle of justice also has implications for the way a trial is designed and carried out.

Consider, for example, a situation in which there are two possible designs for a clinical trial. The first is scientifically a bit more elegant or less expensive to carry out, while the second has a more attractive benefit-to-harm ratio for the subjects. How should a sponsor, an investigator, or an IRB decide which design to operationalize?[12]

I would claim that it is morally relevant whether the subjects are among the worst off members of society. If they are, they have an additional claim (a claim of justice) to arrange the practice so that it is to their advantage. If terminally ill subjects are among the worst off and therefore have this special moral claim of justice, this fact would be relevant in deciding the trade-off between subject well-being and societal interests in maximally efficient research. Many protocols might raise this kind of question about the design and execution of the research.

THE AUTHORITY OF PUBLIC VERSUS PRIVATE CODES OR PRINCIPLE LISTS

One final preliminary problem raised by the *Belmont Report* has received almost no attention. It should be clear that *Belmont* differs substantially from other codifications or lists of moral principles designed to be applied to human subjects research. Nuremberg, if taken literally, prohibits all research on subjects who cannot consent (because consent is the very first requirement of Nuremberg). The most recent statements of the American Medical Association (AMA) authorize experimental treatment without the patient's informed consent in special circumstances — a provision that is almost certainly in violation of the spirit of both Nuremberg and *Belmont*.[13] The American Psychological Association[14] has long endorsed behavioral research on terms that violate both the Department of Health and Human Services (DHHS) regulations[15] and the spirit of *Belmont*.

The issue for now is not which of these codifications is better; it is which source has moral standing. In particular, do publicly generated codifications such as *Belmont* and Nuremberg have authority over privately generated codifications coming from professional organizations or other private groups? It seems clear that the public articulations have standing that cannot be claimed by professional groups. Most subjects cannot even be a member of the professional groups that claim to generate codes for their members. Likewise, virtually all research subjects cannot be members of the organizations behind the Helsinki Declaration. In some cases these professional groups may have developed wise formulations that are persuasive for nonmembers, but these have authority only insofar as their reasoning is persuasive to the nonmembers. No privately generated codification can otherwise claim any authority for justifying human subjects research for subjects who do not voluntarily subscribe to the private documents. That still leaves open the question of whether a document of international public law, such as Nuremberg, should take precedence over a national public body, such as the National Commission, but it at least establishes a clear moral priority for the public over the private documents. I believe subjects should refuse to participate in research under the auspices of any IRB that claims Helsinki or the AMA is the set of moral principles governing its research.

Resolving Conflict among Principles

These five preliminary questions about the *Belmont Report* lend themselves to rather easy resolution. Modest terminological clarifications (such as changing from *beneficence* to *utility*) or specifications (extending justice to research design) or assigning greater moral authority to public documents over professionally generated ones should resolve these. One major underexamined issue, however, is remarkable: that is, the question of whether, for a research protocol to be morally justified, all the principles must be satisfied or whether, when they conflict, the principles taken together must merely be satisfied "on balance."

A REMARKABLE OMISSION IN THE POST-*BELMONT* DISCUSSION

This problem is remarkable because, without addressing it, the *Belmont* principles are virtually meaningless. To be sure, there are some research protocols that are so benign and so easy to execute that all three *Belmont* principles can be fully satisfied simultaneously and easily. There are probably also some research proposals that are so offensive that they not only violate respect for persons and justice but also fail to offer significant benefit. Those proposals, however, will not command much attention from an IRB.

The interesting and controversial cases are those for which significant net benefit is envisioned from the investigation, but the research can only be conducted (or can be conducted most efficiently) if the autonomy of subjects is eclipsed or if justice is compromised by recruiting easy-to-access, low-income patients and prisoners rather than spreading the burden more equitably.

In such cases, it is crucial to know what the moral theory of the National Commission would say. Would it consider all such research morally unacceptable unless it can be brought to satisfy all three principles simultaneously, or would it be sufficient if the benefits to society that are envisioned sufficiently outweigh (whatever that would mean) the infringement of justice and respect for persons? If a research proposal anticipated great benefit, but violated both respect for persons and justice, would it be sufficient to resolve the injustice in subject selection so that the proposal, as modified, would satisfy two of the three principles? If so, would it be acceptable if making the proposal more just reduced the expected net utility from the research? For example, if the additional costs of more equitable subject recruitment meant that the sample size were reduced, thus lowering the probability of a scientifically valid outcome, would it be acceptable to endorse this lower probability of success provided there was still a realistic chance of success? Would "trade-offs" between justice and utility be morally acceptable such that those designing the research should seek out some optimum point where, considered together, justice and utility were maximally satisfied? In such cases, should we also factor in respect for the autonomy of subjects so that less expected utility and less just subject selection could be tolerated to achieve greater autonomy of subjects and vice versa?

These are not mere hypothetical questions that could arise in very special cases of unusual protocols. They are problems that arise in almost all protocols in the real world of human subjects research.

EXAMPLES WHERE THE RANKING/BALANCING CONFLICT IS CRUCIAL

A few examples for contemporary debates over human subjects research will illustrate the problem.

Emergency Room Research. Recent federal regulations permitted emergency room (ER) research without the informed consent of the patient or surrogate in special circumstances.[16] Commentators have long realized that the consent requirement would seem to preclude or at least make very difficult research on unconscious patients in the emergency room. Victims of strokes, head trauma, or accidents may need immediate intervention to attempt to prevent permanent damage. If standard therapies suggest poor prognosis and experimental procedures seem, on theoretical grounds and animal tests, to be quite promising, ER personnel may want to conduct a randomized clinical trial. Such patients cannot give consent, and time may not permit locating valid surrogates. Such patients seem perpetually doomed to poor treatment if we cannot make an exception to the informed-consent rule in such cases. The federal regulations permit such waivers of consent provided the research cannot be conducted otherwise, there is a "community consent," and other conditions are met. As long as there is absolutely no basis for preferring one treatment over the other (we are at indifference[17] or equipoise[18]), we can imagine that no rational person would object to such randomization and that consent could be "constructed" or "implied," thus working our way around the requirement that the subject's autonomy be respected. The problem with this maneuver, however, is that there will come a time in virtually every such trial when a data trend appears to be quite strong but is not yet sufficient to conclude decisively that one treatment is better than the other. This is what I have referred to as the "preliminary data problem."[19]

Consider, for example, when an interim analysis is conducted and one treatment is found superior to the other with a p-value equal to 0.07. That means there are seven chances in one hundred that the difference in the data would have appeared even if there were no difference in the underlying treatments. That is too great a chance to conclude that there is a real difference, but, from the point of view of a patient who must at that moment receive one treatment or the other, the rational preference would be for the apparently better treatment. It would be irrational to presume the patient would be indifferent to a choice between the two options and thus irrational to construct or presume consent to be randomized.

My conclusion is that any such study would fail to satisfy the autonomy component of the respect for persons principle. Then again, if the study were not completed, all future patients would be condemned either to the existing standard treatment or to a treatment that has not been proven superior at an adequate level of significance. We need to know whether we can trade off respect for autonomy in this case when enormous good would seem to follow from doing so.

The problem is even more complicated. In almost all ER research of this kind, it is conceivable that the trial eventually could be done while still requiring either advance consent or surrogate consent. With considerable expense and inconvenience we could try to get advance consent either from a large percentage of the population (using some advance directive mechanism) or from surrogates who happen to accompany the patient to the ER, but the costs would surely increase and the results would be delayed, at the least, if not tainted more directly. We could also interrupt all randomizations when a data trend emerges and continue to collect data, which we would then compare to historical controls, but the results would not be as elegant with these "historical controls." The issue for one who wants to solve this problem by applying the *Belmont* principles is whether we can trade off one principle against the another (compromise respect for autonomy to serve net utility for the society or the other way around) and, if so, how the trade-off should be made.

Desert Shield. A similar problem arose in Desert Shield, the U.S. military effort in the Gulf War. Fearing chemical and biological warfare, U.S. military authorities wished to administer anthrax vaccine and chemotherapeutic agents that were not approved for clinical use under investigational new drug protocols. Moreover, these agents were to be used without the consent of those receiving them.[20]

The forced use of investigational agents by the military poses a wide range of questions, including whether their use constitutes research and whether the purported therapeutic use of an investigational agent controlled by the Food and Drug Administration (FDA) should be governed by the *Belmont* principles and the related DHHS regulations. Because it was the FDA that claimed to authorize waiving consent, one would think that they perceived their regulations, grounded in *Belmont*, to apply. Moreover, it is hard to see why the *Belmont* principles (or a modified version of them) are any less relevant to therapeutic than to research medicine.

There seems to be no question that this FDA waiver, like its waiver for ER research, violates the *Belmont* principle of respect for persons insofar as it requires respect for autonomy. The question is whether these interventions could be justified by *Belmont* through an appeal to net benefits so large that they outweighed the violations of patients' autonomy.

In the case of the ER research, it is easy to imagine that there were potential benefits to society from getting answers to the research questions. However, in the case of the Desert Shield use of investigational agents, it is harder to determine the benefits. The agents had not been approved for these uses. They were still classified as investigational. By definition, the benefits from these uses had not yet been established to the FDA's satisfaction. Yet the only justification for compulsory use would have to be that the decision makers were convinced that the benefits clearly outweighed the harms. It is hard to imagine how large expected net benefits could, under these circumstances, justify violating the *Belmont* principle of respect for persons. Moreover, it is hard to see why consent could not have been obtained. The fact that the troops (and the public) were notified well in advance of their compulsory use suggests that there was time to inform the ones who would be using the agents. Also, to the extent that the benefits seem overwhelming in comparison to the expected risks, very few refusals should be expected. On the other hand, we know that, in principle, military personnel can have their autonomous choices overridden for the benefit of the military mission (i.e., the societal good).

The issue for our purposes is whether, in principle, large amounts of social good — good from accomplishing a military purpose or good from resolving a scientific question — can offset the prima facie reasons based in the other *Belmont* principles to avoid violating autonomy. If each of the principles is merely a prima facie principle that can be overridden by the weighty force of other principles, then "in principle" they can be. On the other hand, if each principle is independently a necessary moral condition that must be satisfied, then neither the emergency room waiver nor the Desert Shield waiver would be justified.

Records Research and Other Risk-Free Research. The problem arises not only when the benefits of the research seem very large and when there may be special benefits from using investigational agents but also in cases in which the risks to subjects are small and when some people believe the slight risks may justify overriding the *Belmont* principles of respect for persons and justice. Certain apparently risk-free research is classified as exempt from DHHS regulations.[21] In fact, the previous generation's Department of Health, Education, and Welfare regulations only applied in cases in which subjects were at risk.[22] This would mean that, if confidentiality were adequately protected, whole classes of research, such as use of body wastes and record searches, could easily be approved without requiring subject consent. If envisioned benefits combined with what is essentially absence of risk can offset requirements based in respect for persons and justice, then there would be no need to notify subjects of their involvement. There would be no need to worry about equitable subject selection, especially if respecting subject autonomy or justice increased the costs and decreased the efficiency of the research. Nevertheless, even if the subjects of such studies can in no way be harmed significantly and the benefits of conducting the research without consent and without attention to equity in subject selection are much greater, there are still reasons why some people want to insist on meeting the requirements of the other principles in these studies. For example, current DHHS regulations require informing subjects of the purpose of the research. The subjects may disapprove of the purpose of the research even if they are not at risk of harm in any way. They may want to exercise their autonomy to refuse to contribute to what they consider an unacceptable purpose. They may object to having their records examined by the researchers even if the researchers guarantee that no further disclosure of their medical information will take place. Real violations of the implications of respect for persons or justice can therefore occur even if the subject is not at risk of harm. If all the *Belmont* criteria must be met for research to be acceptable, the fact that risks are small and benefits are large does not justify overriding this principle.

Conscription of One to Save All. One final case, a purely hypothetical one, is worth considering. Imagine a need to conduct the "ultimate" research project, one necessary to save the world from total destruction; but, in order to do the study, one life would have to be sacrificed. That it is hard to imagine what would require such a project does not negate the challenge the hypothetical situation poses for the *Belmont* principles.

It is possible that someone would become the ultimate volunteer, that he or she would volunteer to be killed for the benefit of saving humankind. (It is an interesting question as to whether asking for and accepting volunteers would be ethical in such a circumstance.) Let us assume, however, that no one volunteered, so the question became one of whether a person could be picked at random for this sacrifice in a fair lottery.

Conscripting someone for this purpose would, of course, violate the autonomy requirements of respect for persons, but it could plausibly meet the requirements of what *Belmont* called beneficence (which I am suggesting should be called utility). The question is whether this ultimate net gain in utility would justify a major breach in the requirements of respect for persons, even for the ultimate good cause. This hypothetical poses most dramatically the question of whether the *Belmont* principles are meant to be fulfilled as independent necessary conditions or fulfilled merely on balance.

Textual Evidence in the *Belmont Report* and Related Documents

Search as I may, I can find no direct answer to this question in the text of the report itself. I think, therefore, we must look for indirect evidence in the *Belmont Report*, the individually authored appendices to the report, other National Commission documents, and the federal regulations that are based on the report. The fact that Commissioner Albert Jonsen has said that the *Belmont* message was "You may not submerge the rights of individuals for the sake of progress," whereas other commissioners clearly implied balancing of individual rights with social interests was the method of *Belmont*.

The appendices to the *Belmont Report*, which at first seem promising as sources of an answer, turn out to provide no definitive information. Two massive volumes of appendices provide twenty-six individually authored chapters written by consultants to the National Commission. These include several written by philosophers who might have been concerned about how the principles should relate to one another and whether all must be satisfied for research protocols to be ethical. However, these chapters have a variety of foci, and, in any event, were never clearly endorsed by the National Commission.

EVIDENCE FOR THE BALANCING INTERPRETATION

The three principles are described in the *Belmont Report* as "general prescriptive judgments" and as providing a "framework that will guide the resolution of ethical problems arising from research involving human subjects." Both descriptions might seem to imply that additional work will be needed by decision makers rather than simply ensuring that each of the three principles is satisfied independently. Moreover, we know that the commission staff philosopher, Tom Beauchamp, is on record favoring a balancing approach in which principles are treated as generating only prima facie duties.

The most convincing evidence in favor of the view that each principle is merely prima facie and can be overridden by the others is in the endorsement of categories of research that are exempt from scrutiny of IRBs. In certain studies (including studies of documents, records, or pathological specimens), if the subjects are not identified or identifiable, the National Commission concluded that the research need not be considered to involve human subjects (and is therefore exempt from the commission's recommendations).[23] The avoidance of risk apparently justifies excluding any requirements based in respect for persons or justice (even though the subjects may nevertheless object to the purpose of the study or find subject selection inequitable). A similar point can be made about the creation of categories of research that are exempt in the DHHS regulations[24] and the subsequent common federal agencies rule.[25] The combination of potential benefit and the absence of risk of harm appears to justify sacrificing respect for persons and justice in these cases.

Still another piece of evidence is in the provision for a waiver of consent in certain low-risk cases provided that several criteria are met. The National Commission recommended that IRBs can approve of withholding or altering information given the subject when disclosure to subjects would "affect the data and the validity of the research."[26] A similar provision appears in the DHHS regulations following the National Commission reports.[27] The fact that there is little or no risk would not justify waiving consent grounded in respect for persons or equitable subject selection grounded in the principle of justice unless these could be traded off against certain net positive utility and the absence of risks of harm.

EVIDENCE FOR A RANKING OR SIMULTANEITY INTERPRETATION

There is also, however, reason to believe that the writers and interpreters of the *Belmont Report* do not always see the principles as being traded off against one another. The provisions for a consent waiver specify that, in addition to the conditions mentioned earlier, consent cannot be waived unless "sufficient information will be disclosed to give subjects a fair opportunity to decide whether they want to participate in the research."[28] Similarly, the DHHS regulations require that, to waive consent, "the waiver will not adversely affect the rights and welfare of the subject."[29] This generates confusion, however, because at least some people understand that one of the rights that cannot be violated is the right of the subject to give an adequately informed consent. Even if that is not one of the rights, the general notion that consent cannot be waived unless rights are protected implies that there is such a thing as rights and these rights cannot be offset by the interests that would be served by conducting such research without consent. The implication seems to be that subjects possess certain rights (perhaps such as the right to consent to the purpose of the research and the right to be selected equitably) that cannot be offset by anticipated great good and absence of risk of harm.

An additional National Commission recommendation that seems inconsistent with a balancing approach is its attitude about research that could cause serious harm to unconsenting subjects. The commission struggled mightily to develop language that would endorse research involving an intervention that does not hold out the prospect of direct benefit for the subject and involves more than minimal risk (say, in its report on children).[30] It finally developed compromise language, but stipulated that such research would be limited to cases where "such risk represents a minor increase over minimal risk" or that the research "presents an opportunity to understand, prevent, or alleviate a serious problem affecting the welfare of children." For our purposes, the key question is why there is an absolute prohibition on risks beyond this level when the welfare of children will not be served. If a balancing approach were used, the proper approach would have been to make all the appropriate statements about how there must be a strong presumption against going beyond minor increases over minimal risk, but then to concede that, if large quantities of beneficence for adults were at stake, proportionally larger risks could theoretically be justified. Keeping in mind that the National Commission treated protection of children from harm as a requirement based on the principle of respect for persons, it seems that the commission is saying that respect for persons places an absolute limit on beneficence. No matter how much benefit for adults is envisioned from risking harm to a subject who is a child, that child has a right (grounded in respect for persons according to the National Commission) to be protected. No balancing of beneficence for adults against respect for persons is tolerated, at least when the risk is more than a minor increase over minimal.

I am not aware that the National Commission ever confronted a similar problem regarding justice. A balancing approach would seem to imply that compromises with justice could be tolerated if the beneficence at stake were only large enough. However, *Belmont* never concedes that justice can be compromised in such cases. It seems that the National Commission must have intended that justice and respect for persons pose necessary conditions for research to be ethical, not prima facie norms that can be sacrificed in the name of beneficence if only the expected benefits are large enough.

A similar question can be asked regarding the DHHS and common federal rules. The key sentence in the regulations states, "In order to approve research covered by these regulations the IRB shall determine that *all* of the following requirements are satisfied" [italics added].[31] There follows a list of the famous seven criteria, including those that seem clearly grounded in the *Belmont* principles of respect for persons and justice. These include that, in addition to consideration of overall benefit, the risks to the subject are "minimal" and "reasonable," that "selection of subjects is equitable," and that "informed consent will be sought from each prospective subject." What is important is that *all* seven criteria must be met. The text does not permit them to be met "on balance." There can be no case in which sufficiently weighty amounts of beneficence would justify omitting consent or equitable subject selection. Thus there are provisions in the National Commission's work and the successor documents that seem to imply a balancing of conflicting principles, but also some provisions that seem to make no sense unless respect for persons and justice were independent necessary conditions for research to be ethical, conditions that could not be offset by balancing them against beneficence.

The Need for a Theory

The ambiguity in the *Belmont Report*, the appendices, the other National Commission documents, and the government regulations following from the commission's work suggests that there is an urgent need to develop a theory of what should happen when one or more of the principles is maximally satisfied by one set of research design decisions while another principle requires some other set of decisions. These might be overarching IRB decisions whether to endorse a protocol, or they may be more subtle decisions about choosing one variation in alternative research designs.

SINGLE PRINCIPLE APPROACHES

One theoretical stance would be to affirm one single principle as foundational. Utilitarians, for example, affirm the principle of utility (presumably the National Commission's beneficence) as the definitive principle. Presumably, a utilitarian investigator or IRB would endorse a protocol that maximized net anticipated good consequences even if that meant ignoring considerations of respect for persons or justice. These other two principles would function, as the National Commission sometimes suggests, as mere guidelines, as warning flares that utility might not be maximized in cases in which justice in subject selection or subject self-determination was ignored. Rule utilitarians might give the other principles prominent place, letting them function as rules that are not normally violated in the individual case. Rule utilitarians, however, would hold open the possibility that the rules might have to be specified differently if it turns out that some other formulation was actually utility maximizing. They would presumably be open to exceptions to the consent rule in cases in which the research could not feasibly be conducted if consent were required or in cases in which subjects were at essentially no risk. At least at the level of rule specification, the only truly operative principle would be utility.

Likewise, a libertarian might elevate respect for subject autonomy to the highest position using utility and justice as signals of when liberty might be in jeopardy. It is, however, much harder to make the case that utility and justice are rules that tend to ensure liberty in the normal situation. The true libertarian would be more likely simply to invoke liberty as a single principle, prohibiting all research that does not respect persons' autonomy. Similarly, a true egalitarian would endorse a single principle of justice as the sine qua non of ethical research. It seems quite clear that the National Commission was not grounding its position in a single principle and that it would be hard for anyone else to make such a case.

BALANCING STRATEGIES

A much more plausible case can be made for the view that the National Commission must have intended the *Belmont* principles to be balanced against one another so that, on balance, the three principles were satisfied in the best possible way (even if one — or even two — of the principles was set aside or outweighed in a final determination that the research was ethically acceptable). Such a balancing — or a conflicting appeals approach, to use Brody's phrase — is quite appealing, at least at first glance. A more careful examination, however, reveals that such balancing (or intuitional resolving of conflict) has very troublesome implications. It means that, if only the consequences are good enough, it is acceptable to omit respect for persons and the consent it requires. At the very least it seems that such omissions violate the direct meaning of the Nuremberg code when it says that "the voluntary consent of the human subject is absolutely essential." It means that if the National Commission principles are to be balanced, the commission was in violation of the explicit words of Nuremberg. It also raises the question of why, if the consent requirement implied in respect for persons can be balanced away in the name of sufficient utility, there has never been a civilian American legal case since the release of the *Belmont Report* in which large amounts of good consequences have been seen as justifying omission of consent.

It also means that, if sufficiently good consequences permit the waiving of the implications of the principle of justice, there would be exceptional cases in which utility would justify occasional research exclusively on prisoners, low-income groups, or charity patients even though better-off subjects with more freedom could have, with greater cost, been recruited. It would mean that if there were sufficient utility in limiting subjects to Caucasians or males, doing so would be in accord with *Belmont*, even though other racial, ethnic, or gender groups could, at greater cost, have been recruited.

It is striking that essentially without exception we now insist that utility can never justify ignoring respect for the choices of substantially autonomous persons to refuse to participate in research and can never justify inequities in subject selection.

It is possible — though it seems unlikely — that the seemingly absolute requirement that autonomy and justice be satisfied for a protocol to be ethical is merely a rule-utilitarian way of resolving the conflict between utility and the other principles. A far more plausible explanation would be that respect for persons (with its implications for consent, confidentiality, veracity, and avoidance of intentional killing) and justice are fundamental requirements of morality that cannot be overcome by piling on more and more expected utility. They are each necessary moral conditions that must simultaneously be satisfied in every protocol. Conversely, if the principle of utility is taken to require utility maximizing in order for it to be satisfied, it is obvious that we sacrifice utility all the time in order to make protocols more equitable and to make them show more respect for persons. Certain research is simply unethical in its inception regardless of the fact that good can be envisioned in the results. Other research is conducted in a manner that is expected to be beneficial even though it could be even more beneficial if respect for persons and justice were overruled. For example, even though the results would be more reliable if no subject approached to be in a study were permitted to decline, we nevertheless insist on the right of potential subjects to refuse to consent.

RANKING STRATEGIES

This suggests that the proper resolution to the question of what should happen when the *Belmont* principles come into conflict is to rank the principles with utility being subordinated to other considerations.[32] The principle of utility would thereby be interpreted to require the maximizing of utility consistent with the prior constraints of the principles of autonomy and justice. This approach poses a problem, however. Most people intuitively believe that no definitive and complete ranking of principles can be sustained. They believe that eventually there will be a set of circumstances in which any one principle will have to win out in order to square with our firmly held considered moral judgments.

One such argument focuses on our extreme hypothetical situation in which conscripting one unwilling subject for an experiment that will cause death is necessary to save the entire world. Before looking at that difficult hypothetical, however, we should consider the intuition that most people find it extremely difficult to rank respect for persons uniformly over justice or the other way around.

I think the only reasonable response to this intuition is to concede that respect for persons and justice are extremely important moral considerations and that, if they conflict with one another, there are times when we will want to sacrifice autonomy to promote justice (e.g., in some cases of compulsory taxation and military conscription) and some other cases in which we tolerate some injustice in the name of respecting autonomy. If, as I have suggested, respect for persons is more complicated and involves more than merely respecting persons' autonomy—if it involves veracity, fidelity to commitments, and avoidance of killing as well—a similar point can be made.

There are cases in which each of these may have to be sacrificed in the name of justice and cases in which justice may have to be sacrificed in the name of each aspect of respect for persons. In fact, it seems more and more that justice and autonomy can be conceptualized as just more specific principles of the overarching super-principle of respect for persons. It is merely another aspect of respect for persons in which we hold that persons should be treated equitably.

In that case it makes no difference whether respect for persons is considered one principle or several. The different elements that make it up—respect for autonomy, veracity, fidelity, avoidance of killing, and justice—must be balanced against each other to lead to an overarching moral requirement that has priority over mere utility maximizing.

MIXED RANKING AND BALANCING

This suggests that the proper way of resolving conflict among the three *Belmont* principles involves a combination of ranking and balancing. We must first balance the components of respect for persons as well as justice (if justice is thought to be distinct from respect for persons) to determine what these principles that do not strive to maximize good consequences call for. Likewise, we must balance beneficence and nonmaleficence to determine what course is utility maximizing. Once these two preliminary steps are completed, I suggest we can only get our theory to square with our considered moral judgments if we give respect for persons and justice—taken together—absolute priority over the utility-maximizing principles of beneficence and nonmaleficence. That is, a research protocol is justified only if, on balance, it respects subjects and conforms to the requirements of justice. Within that constraint, it must then be determined that more good than harm is expected. Furthermore, among all the possible protocols that conform to the requirement of elements of respect for persons balanced with justice, the protocol that entails the greatest net good is most acceptable. In this sense, respect for persons, justice, and utility must be satisfied simultaneously. The *simultaneity view* wins out. If no research design offers any net positive utility from among those that meet the test of respect for persons balanced against justice, then no research in this area may be conducted.

We are left with the great hypothetical: the case where we believe that it is necessary to conscript one unwilling subject and condemn that subject to death in order to save all of humankind. This would appear to be the ultimate case in which one might want to sacrifice respect for a person in order to do great good. I speculate that many people believe such conscription would be morally acceptable if there were no alternative, so I need to show how the principles under the rubric of respect for persons (including the justice principle) collectively take priority over utility.

I must first note that, like all hypotheticals, such an extreme case is unlikely to occur in reality. For example, before a society is forced to conscript someone for this role, one or more volunteers may step forward. Many societies have permitted such altruism (the Christian martyrs, Japanese kamikaze pilots, volunteers for military service, even volunteers for self-experimentation with smallpox and yellow fever, etc.). The question remains, however, of whether it would be acceptable to conscript one subject if necessary to maximize social benefit.

In order to reply we must differentiate between conscription (i.e., violation of autonomy) in the name of aggregate social utility and conscription to serve some other nonconsequentialist principle. It is possible that the one conscripted may, in addition to contributing to the greater aggregate social good, also increase justice in the world by benefiting those who have the greatest claims of justice. For example, many theories of justice hold that justice requires arranging social practices to benefit the least well off or make people more equal. Violating persons' autonomy may sometimes benefit the least well off and thereby contribute to justice according to this understanding of that principle. If, however, autonomy and justice must be balanced against one another, as I have suggested, then some violations of autonomy would be in order for some promotions of justice. The conscription of one subject would, in addition to maximizing social utility, also contribute to the welfare of the least well off. Thus, we need to know whether it is the aggregate social utility or the promotion of justice to the least well off that justifies infringements on autonomy. It could be that it is the latter and not the former. If so, even if we are convinced that, in some extreme cases, sacrificing autonomy of research subjects is justified, it could be that it is justice or some other aspect of respect for persons that provides the justification, not mere utility maximization.

This would provide a very small window to overrule the requirement of respect for autonomy. Subject autonomy could be balanced against justice even if it could not be balanced against utility maximizing. Such a resolution of the problem of conflict among ethical principles is the proper one. Not just any old consideration of large amounts of utility would justify suppressing the principle of respect for a person's autonomy. In fact no consideration of utility maximizing per se would. Justice and the other dimensions of respect for persons could, in theory, provide a basis for suppressing the principle of respect for autonomy. Only when, on balance, these other components of the requirement of respect for persons outweigh respect for autonomy would autonomy be defeated. Only when respect for persons — the balanced consideration of autonomy, veracity, fidelity, avoidance of killing, and justice — is fully satisfied would utility — beneficence and nonmaleficence — come into play. The former are first balanced and then collectively ranked against utility (or what *Belmont* calls beneficence). In this sense, respect for persons and a subordinated version of utility must be satisfied simultaneously (as must justice, insofar as it is a separate principle). *Belmont* never tells us exactly what should happen when its principles come into conflict, but I believe this is the most plausible answer. The only alternative is to admit that the Nuremberg principle of self-determination can be violated and that unjust subject selection and research design are acceptable only if the social payoff is large enough.

Notes

1. National Commission for the Protection of Human Subjects of Biomedical and Behavioral Research, *The Belmont Report: Ethical Principles and Guidelines for the Protection of Human Subjects of Research* (Washington, DC: Government Printing Office, 1978).

2. See Tom L. Beauchamp and James F. Childress, *Principles of Biomedical Ethics*, 5th ed. (New York: Oxford University Press, 2001); Raanan Gillon, ed., *Principles of Health Care Ethics* (New York: Wiley, 1994); Edmund D. Pellegrino and David C. Thomasma, *For the Patient's Good: The Restoration of Beneficence in Health Care* (New York: Oxford University Press, 1988); H. Tristram Engelhardt, *The Foundations of Bioethics*, 2nd ed. (New York: Oxford University Press, 1996); and Robert M. Veatch, *A Theory of Medical Ethics* (New York: Basic Books, 1981), for examples of individuals who, even though they manifest significant differences in their theories, I would place in the principlist camp. For critiques, see Ronald Green, Bernard Gert, and K. Danner Clouser, "The Method of Public Morality Versus the Method of Principlism," *Journal of Medicine and Philosophy* 18, no. 5 (1993): 477–89; and K. Danner Clouser, "Common Morality as an Alternative to Principlism," *Kennedy Institute of Ethics Journal* 5 (1995): 219–36.

3. Bernard Gert, *Morality: A New Justification of the Moral Rules* (New York: Oxford University Press, 1988).

4. Albert R. Jonsen and Stephen Toulmin, *The Abuse of Casuistry: A History of Moral Reasoning* (Berkeley: University of California Press, 1988).

5. Both more traditional virtue theories (including those of secular as well as religious origins) and more contemporary feminist theories, including care theories, come to mind.

6. Baruch Brody, *Life and Death Decision Making* (New York: Oxford University Press, 1988).

7. W. D. Ross, *The Right and the Good* (Oxford: Oxford University Press, 1930), 21.

8. Gert, *Morality*, 96–159.

9. Veatch, *A Theory of Medical Ethics*.

10. This is the view of Ross, *The Right and the Good*, 21, who called it a different duty from beneficence and "of a more stringent character." This was also the view of Childress in the early editions of *The Principles of Biomedical Ethics*.

11. Robert M. Veatch, "Justice and Research Design: The Case for a Semi-Randomization Clinical Trial," *Clinical Research* 31 (February 1983): 12–22.

12. I once monitored a study that proposed to administer a toxic chemotherapeutic agent to terminally ill cancer patients and reverse the toxic effects with a "rescue agent." The investigators wanted to administer the drugs by hospitalizing the patients, but that would require having patients with a short life expectancy in the hospital for three of every twenty-one days. The drugs could be

administered at home with greater convenience for the subjects, but that design was believed to be somewhat weaker science (because the patient's environment would be less controlled) and also more expensive (because nurses would have to be sent to the homes to administer the medication). The issue was which design is morally preferable.

13. American Medical Association, Council on Ethical and Judicial Affairs, *Code of Medical Ethics: Current Opinions with Annotations, 1998-1999 Edition* (Chicago: American Medical Association, 1998), 17–18.

14. American Psychological Association, "Ethical Principles of Psychologists and Code of Conduct" (1992), as reprinted in *Encyclopedia of Bioethics*, rev. ed., vol. 5, ed. Warren T. Reich (New York: Simon & Schuster Macmillan, 1995), 2743–53. See especially p. 2751, article 6.15, in which deception in research is explicitly condoned in violation of the *Belmont* principle of respect for autonomy and without even the precautions stated in the DHHS regulations requiring that the rights of subjects be protected if consent is waived.

15. U.S. Department of Health and Human Services, "Federal Policy for the Protection of Human Subjects: Notices and Rules," *Federal Register* 46, no. 117 (June 18, 1991): 28001–32.

16. United States Department of Health and Human Services, Food and Drug Administration, "Protection of Human Subjects; Informed Consent and Waiver of Informed Consent Requirements in Certain Emergency Research: Final Rules," 45 C.F.R. 46, *Federal Register* 61, no. 192 (October 2, 1996): 51498–533.

17. Robert M. Veatch, *The Patient as Partner: A Theory of Human-Experimentation Ethics* (Bloomington: Indiana University Press, 1987), 210–11.

18. Benjamin Freedman, "Equipoise and the Ethics of Clinical Research," *New England Journal of Medicine* 317 (1987): 141–45.

19. Robert M. Veatch, "Longitudinal Studies, Sequential Design, and Grant Renewals: What to Do with Preliminary Data," *IRB* 1 (June/July 1979): 1–3.

20. United States Department of Health and Human Services, Food and Drug Administration, "Informed Consent for Human Drugs and Biologics: Determination That Informed Consent Is Not Feasible: Interim Rule and Opportunity for Public Comment," *Federal Register* 55, no. 246 (December 21, 1990): 52814–17; Curt Suplee, "FDA Consents to Use of Unapproved Drugs on U.S. Desert Troops," *Washington Post*, December 22, 1990, A10; Guy Gugliotta, "U.S. Forces in Gulf Notified of Vaccinations against Anthrax," *Washington Post*, December 31, 1990, A8.

21. DHHS, "Federal Policy for the Protection of Human Subjects," 28012–13.

22. Department of Health, Education, and Welfare, "Protection of Human Subjects," *Federal Register* 39 (May 30, 1974): part 2, 18917.

23. National Commission for the Protection of Human Subjects of Biomedical and Behavioral Research, *Report and Recommendations: Institutional Review Boards* (Bethesda, MD: DHEW Publication No. [OS] 78-0008, 1978), 29.

24. U.S. Department of Health and Human Services, "Final Regulations

Amending Basic HHS Policy for the Protection of Human Research Subjects; Final Rule: 45 CFR 46," *Federal Register: Rules and Regulations* 46, no. 16 (January 26, 1981): 8386.

25. DHHS, "Federal Policy for the Protection of Human Subjects," 28012–13.

26. National Commission, *Institutional Review Boards*, 26.

27. DHHS, "Final Regulations Amending Basic HHS Policy for the Protection of Human Research Subjects," 8390.

28. National Commission, *Institutional Review Boards*, 26–27.

29. DHHS, "Final Regulations Amending Basic HHS Policy for the Protection of Human Research Subjects," 8390.

30. National Commission for the Protection of Human Subjects of Biomedical and Behavioral Research, *Research Involving Children: Report and Recommendations* (Washington, DC: Government Printing Office, 1977), 7–8.

31. DHHS, "Final Regulations Amending Basic HHS Policy for the Protection of Human Research Subjects," 8389.

32. According to some theories of the relation between beneficence and nonmaleficence, a similar rank-ordering is called for between these two consequence-oriented principles. Nonmaleficence may take moral priority over beneficence according to these views. I have, however, suggested that it is preferable to view beneficence and nonmaleficence on par with the other, with neither holding a special ranking.

14 Specifying, Balancing, and Interpreting Bioethical Principles

Henry S. Richardson

A number of theorists writing on bioethics have determined that norms may be progressively specified, an idea that I articulated, drawing on Aristotle and Aquinas, in an article published fifteen years ago.[1] I seek now to argue that the lessons of that article, and the potential of the idea of specifying norms, have not been fully assimilated even by the writers on bioethics who have taken it up.[2] In particular, I want to argue against the continued reliance of the bioethics literature on the metaphor of balancing, a metaphor that specification can and should displace.

In a nutshell, my complaint is that relying on the metaphor of balancing leads one to offer the mere semblance of reason giving, where real reason giving is wanted, and the mere appearance of guiding action, where actual guidance is wanted.

But *are* reason giving and action guidance really wanted? They should be. Let me make explicit three relatively weak and noncontroversial premises that imply that they should be wanted in the context of bioethics. First, I suppose that — in possible contrast to general ethical theories, which may confine themselves to ruminating on truth conditions — bioethical theory is meant to be action guiding, either by itself or in combination with some body of commonsense material.[3] While it may have a role in articulating understandings and crystallizing awareness, it also intends to guide action in the future. Second, I suppose that bioethical theory ought to guide action by offering reasons or justifications for certain courses of action and that this justification must meet a weak requirement of publicity. Justifications must be offered in terms of reasons that may be publicly stated. While my argument depends only on *potential* public expressibility of reasons, one might well believe in a stronger requirement that the reasons *actually* be publicly expressed or provided, at least in the context of developing normative constraints on public policy, which bioethics does. Third, I suppose that developing an adequate set of action-guiding principles — at least in a fast-changing context such as bioethics, if not in human life in general — requires the progressive collaboration of many practitioners and theorists, each building on the work of others. These basic premises will be the basis of my argument that the residual place balancing holds in bioethical theory should be ceded to the alternative idea of specifying or interpreting norms.

Before I explain the relation among specification and other modes of interpretation, let me distinguish the different residual roles in which balancing is cast. Balancing may enter either (1) as a feature or implication of the content of a theory's principles or (2) as part of what the theory says about how conflicts among its principles are to be dealt with. In the first possibility, balancing may either be (1a) piecemeal and contextual, or it may be (1b) more global or overall. An example of piecemeal balancing is that, invited by a rule of rescue according to which, if one is in a unique position to help someone who urgently needs help, one ought to provide the help whenever one can do so at relatively minimal cost to oneself. This principle invites balancing in each context, in that what seems "minimal" in relation to saving someone's life will not necessarily seem "minimal" in relation to preventing them from ruining their suit. Piecemeal and contextual balancing of this kind, as dictated by the content of some principle, is relatively innocuous, unobjectionable, and ineliminable and is not what I argue against.

Balancing as a mode of conflict resolution (2) and global or overall balancing (1b), by contrast, are far more troubling.[4] In this chapter, I examine examples of each. My example of a bioethical theory that, while somewhat open to the importance of specifying and otherwise interpreting norms nonetheless adheres to an objectionable extent to the idea of global or overall balancing of type 1b, is the account offered by Bernard Gert, Charles Culver, and K. Danner Clouser in *Bioethics: A Return to Fundamentals*.[5] A theory that, despite a generous endorsement of the importance of progressively specifying principles, persists in relying importantly on type 2 balancing in prescribing how to resolve conflicts among principles is that put forward in the fourth edition of Beauchamp and Childress's *Principles of Biomedical Ethics*.[6]

I have no rival bioethical theory to offer in place of either of these. I am not a bioethicist, and I defer (on principled grounds[7]) to those who are knowledgeable about these topics. My target is not one of the overall theories of either set of authors but their use of the idea of balancing ethical principles against each other. Specification, I argue, offers a fruitful alternative for those developing bioethical theory. It offers, not an alternative *theory*, but an alternative way to conceive of the relationship between a theory's norms and the guidance of action.

Note that by "developing bioethical theory," I do not necessarily mean developing or defending principles peculiar to bioethics. My argument here leaves open the possibility that all true and useful principles of bioethics are counterparts of more generally applicable principles — and indeed, the notion of specification explains how one principle can stand as a counterpart of another. Hence, my argument leaves open whether or not, when one is soundly developing a set of action-guiding principles for bioethics, these

principles are "new" in any meaningful sense. By developing bioethical theory, I mean to embrace whatever one might plausibly be doing in an original
book about bioethics or biomedical ethics.

I begin by clarifying the idea of specification and, in particular, its relation to other modes of interpretation. After next comparing the views of
Gert, Culver, and Clouser with those of Beauchamp and Childress, high-
lighting their different treatments of balancing, I examine each of them in
turn. I close with a discussion of the acceptable place for piecemeal, contextual (1a) balancing as a feature or implication of the content of norms.

Specification and Other Modes of Interpretation

The aim of my article "Specifying Norms" was to displace two leading models of how to bring to bear norms on cases: the model of application (or
deductive subsumption) and the model of balancing. Far from being intended
as a complete moral theory, the model of specification presupposed that one
had a theory, or at least an articulated set of norms, already in hand. A question that then arises is how to bring these norms to bear in guiding action. In
my original essay, I concentrated on the form this problem takes when two
norms conflict; but the issue of how to think about bringing norms to bear on
cases at hand — and the usefulness of the notion of specification — is not limited to conflict cases. One way to bring norms to bear is by deductively subsuming a case under a rule; but this demands, unrealistically, that norms be
universal generalizations in their logical form.[8] Another way is by situational
or perceptive intuition (as in one version of *phronêsis*), but this leaves the reasons for the decision unarticulated. Note that what I have just described are
ways of bringing norms to bear on concrete situations, not specifically ways
of dealing with conflicts among norms. Specification provides an alternative
that can work with norms that are not universal generalizations and can help
us articulate our reasons in a potentially public way.

The principal rivals of the idea of specification, then, are the ideas of
application (or deductive subsumption) and of balancing. Each of these ideas
floats relatively free from the content of moral theory. In judging their relative merits, they must be combined with some set of norms, filling out some
ethical theory or other.[9] The considerations I will be advancing here, however, are largely independent of the content of moral theory. I am concerned
with different ideas about how one might bring moral theory to bear in set-
tling concrete issues.[10]

If we shift the focus to situations in which two principles or other norms
conflict, specification again offers an overlooked alternative. Intuition here
becomes intuitive balancing, an intuitive sense of which set of considerations
"weighs" more. Corresponding to deductive subsumption is an approach that
would seek a fixed way lexically or otherwise to rank principles. Both of these
approaches assume that the set of principles remain fixed and must either
already contain a priority rule that resolves the question at hand or be sup-

plemented by intuitive balancing to determine which ones override. Central to the role of specification, however, is that it is a relation between two norms: an initial one and a more specific one that is brought to bear on practice. Specification, then, can sometimes resolve conflicts by filling out — and thereby changing, at least by addition — the set of norms. Robert Veatch has appeared to deny this third possibility. He writes that "the only other possibility for resolving conflicts among principles [aside from balancing them] is to attempt to rank-order them." When he later comes to a discussion of specification, he assimilates it in these two other possibilities via an assumption that specification will be possible only when the two principles in conflict may be lexically ranked.[11] This ignores, however, that a specification may be quite context specific, and so not even speak to broad aspects of the initial principle's domain, let alone take a position on its relative ranking within those areas.[12]

Specification is not the only mode of modifying principles. In the remainder of this section, I seek to distinguish it from some of the others. Let us assume, together both with Gert, Culver, and Clouser and with Beauchamp and Childress, that the main work that needs to be done in order to achieve progress in bioethics at this time is developing greater concreteness. The work of *abstraction* — crystallizing a few broad principles out of a mass of messier materials — has already been done by other philosophers and bioethicists.[13] Even in the direction of greater concreteness, however, not all moves count as specifications, in my sense. It will, I hope, be useful and interesting to distinguish some of the possibilities.

To begin with, we should distinguish between interpreting norms and deriving subordinate norms. Subordinate norms may be derived from initial norms either by deductive subsumption or by less formal causal reasoning. An example of the former kind of derivation would be deriving "do not lie" from "do not do anything with the intent to deceive," plus a definition of lying as stating a falsehood with the intent to deceive. An example of a more informal derivation in Gert, Culver, and Clouser is their derivation of "do not drink and drive" from "do not kill."[14] Given causal facts about automobiles, alcohol, human bodies, and the general practice of sharing the roads, drinking and driving poses too high a likelihood of killing to be acceptable. While the prohibition of drunken driving is an important and firm one, there is no need to regard it as *interpreting* the rule against killing. If the relevant causal facts changed sufficiently, we would change our views about drinking and driving. The moral prohibition against drunken driving is thus *derivative* from a prohibition on killing (or on imposing undue risk of injury or death). The initial prohibition itself remains unmodified by the derivation, which merely links it to a conclusion by causal (or conceptual) facts. These links supplement the initial norm without changing it. By contrast, an interpretation, as I'll be using the term, *modifies* the content of a norm.[15]

Specifying norms is but one of at least four modes of interpreting them.[16] To be able to discuss these possibilities, I need to mention some further details regarding my analysis of the specification relation. When I initially defined specification, I was concerned in the first instance with the question of how an initial norm may be brought to bear upon a situation. Accordingly, it was important to say precisely in what relation the more specific norm stood to the initial norm, such that we might be licensed in saying that the initial norm is brought to bear when we use the more specific norm. In order to capture this feature, I invoked the semantic condition of *extensional narrowing*: that is, everything that satisfies the specified norm must also satisfy the initial norm, or, if the norms are not logically absolute, everything that satisfied the "absolute counterpart" of the one must satisfy the absolute counterpart of the other.[17] Yet, to fit the intuitive notion of specification, not every norm that happens to be narrower can count. Accordingly, I added a syntactic condition of *glossing the determinables*. That is, I required that a specification narrow a norm by adding clauses spelling out where, when, why, how, by what means, to whom, or by whom the action is to be done or avoided. You can see from this explanation that there are two kinds of interpretation that do not count as specification but are interesting and important nonetheless — namely, moves that simply narrow without glossing and moves that gloss without narrowing. An example of *narrowing* that does not gloss is the move from "do not harm" to "do not torture." The latter is indeed extensionally narrower, but because "torture" is a well-understood notion on its own, there is no need to generate this more specific norm by adding clauses to the initial norm prohibiting harming. An example of *glossing* that does not narrow would be any gloss that purports to replace an initial formulation by defining it rather than supplementing it. For instance, "do not have sex in the office" could be glossed as "do not have sex in the office, by which we mean: do not engage in any act involving contact (of a certain kind) with the genitals of another." This formulation adds words but purports, at least, not to narrow: it is simply spelling out what "having sex" already meant.

Or do we understand sex so definitely? So far, I have been speaking as if the condition of extensional narrowing were noncontroversially applicable — as if we knew in general both what it was for norms of various types to be satisfied and when it is that they are. There are, however, both general problems about the former (which I will not go into here) and frequent cases in which our norms are so vague that it is indeterminate whether or not a given type of action would satisfy them. I owe to my colleague Mark Murphy the idea that *sharpening* a vague norm cannot count as a specification because the specification relation depends on the extensional narrowing condition, and whether the narrowing condition is met will be in principle indeterminate if the initial norm is sufficiently vague (as I had been assuming, in the last paragraph, that the prohibition on having sex was not). For example, neither "do not drink more than the civil law allows" nor "do not drink more than twelve beers at a single sitting" can count as specifications of "do not drink inordinately" — not only because each drops the word *inordinately* but also because it is indeterminate whether every action that satisfies the latter will satisfy the former. Perhaps drinking eleven beers at a single sitting, or a little less than the civil law allows, is also inordinate. The idea of "drinking inordinately" is too vague to determine whether the proposed interpretation counts as a specification of it or not.

Often we are in this position. Beauchamp and Childress write that "a typical example of a rule that specifies the principle of respect for autonomy by giving it more content is, 'Follow a patient's advance directive whenever it is clear and relevant.'"[18] Strictly speaking, in the terminology that I am now suggesting, this is a sharpening that may or may not narrow, but it is certainly not a specification. It does not proceed simply by adding clauses to the principle of respect for autonomy. Further, whether all cases of following such advance directives are cases of respecting autonomy may well be, because of the vagueness of the latter notion, too indeterminate to settle[19] (no more indeterminate, perhaps, than a similar judgment involving the notion of freedom, but indeterminate nonetheless). The connection that Beauchamp and Childress identify between respecting autonomy and following advance directives is a theoretical achievement that takes a relatively complex argument to set out. To be sure, *once* that connection has been made, it will be plausible to state the resulting principle as "respect the autonomy of patients by following their advance directives whenever they are clear or relevant." This *does* count as a specification of the initial norm. All I am insisting here is that, by definition, a specification wears on its face its connection back to the initial norm, for my requirement that specifications proceed by adding clauses implies (or was intended to imply) that the initial clauses will remain.[20] In this way, "respect the autonomy of patients" bears a transparent relation back to "respect autonomy." In the absence of such a transparent link, what Beauchamp and Childress put forward is technically an interpretation of autonomy that sharpens it in a certain way for a certain context rather than a specification of the principle of autonomy. I have distinguished four modes of interpretation that move in the opposite direction of abstraction: specification itself; extensional narrowing and glossing; the components of specification; and sharpening, a prerequisite of specification. Having this nonexhaustive list in mind will be helpful in the following discussion of the ways in which the two sets of bioethics authors take up the idea of specification and, more generally, acknowledge the need for interpreting principles.

Comparing Two Views

The respective approaches of Beauchamp and Childress and of Gert, Culver, and Clouser differ saliently in one major respect. It is a difference that emerges from a background of considerable similarity and complementarity — a background that is important to discuss so that we may have the outlines of their views before us.[21] To be sure, my focus is on the abstract structure of their views, hardly at all on the specific content, let alone on content specific to bioethics. Let me start by mentioning five important structural similarities.

First, both sets of authors put forward *a relatively small number of central principles or rules* that they draw from "our common morality"[22] and that they use to help generate more concrete guides to action. Beauchamp and Childress offer four: autonomy, justice, beneficence, and nonmaleficence. Gert, Culver, and Clouser have ten. They factor nonmaleficence into five rules, each directed against a particular kind of harm, adding prohibitions on deceiving, breaking promises, cheating, breaking the law, and not doing one's duty.[23] I am less interested in the specific lists than in the fact that each group of authors lists a finite number of independent moral principles or rules.

Second, both sets of authors characterize their rules or principles as representing what W. D. Ross called "prima facie duties." That is, each of the duties they list is subject to being overridden by other moral considerations. Further, both sets of authors describe this process of norms being overridden in a way that invokes the metaphor of weighing or balancing. Beauchamp and Childress align themselves more explicitly and directly with Ross in this respect, putting the point in the quantitative language of balancing: each of the principles holds "unless it conflicts on particular occasion with an equal or stronger obligation."[24] Gert, Culver, and Clouser do not directly state that it takes a moral consideration to override a moral rule and do not speak of objectively "stronger" obligations; instead, they elaborate a hypothetical standard for when rule violations may be justified. According to them, a moral rule may be violated only if "an impartial rational person [could] advocate that violating it be publicly allowed."[25] Rationality here requires that the person advocating the violation have an "adequate reason" in mind, as judged by "some significant group."[26] Impartiality requires not being influenced by which persons are affected by the violations.[27] Although this combination of rationality and impartiality does not entail that only moral reasons (or what some significant group takes to be such) may override a moral rule, they do tend in that direction.

Third, both sets of authors take an important step beyond Ross's model of balancing conflicting prima facie duties by insisting on the importance of *interpreting* the moral principles, in the sense developed in the last section of this chapter. Both stress that these interpretations must take account of existing social practices without kowtowing to them. Both indicate that it will be necessary to adapt the general principles from which they start so as to fit the special requirements of the biomedical context.[28]

Fourth, both sets of authors indicate that there will be a *need to specify* the principles or norms in the process of interpreting them. In general, their calls for interpretation reflect an awareness of the highly general character of the principles and rules that they invoke and of the gap that therefore arises between them and the guidance of action in particular circumstances. The interpretations aim to make the principles more specific in ways that take account of concrete aspects of the biomedical context. In Beauchamp and Childress, this call for specification is pervasive.[29] Gert, Culver, and Clouser stress instead other modes of interpretation, but do take comfort in their rivals' adoption of specification, for they think that well specifying the four principles will mark at least a "way station" on the journey to truth — that is, toward their own alternative view of the moral rules governing bioethics.[30] They distinguish between what they call "general" moral rules, which all rational persons have reason to follow, and "particular" moral rules, which are contextual specifications and interpretations of the general rules.[31] General moral rules, they note, are so abstract that they need "culturally sensitive specification" to be usable at all.[32] To be culturally sensitive, this specification will necessarily have to be different in different communities.

Fifth, both sets of authors nonetheless claim that there will be an important and necessary place in the theory for the idea of *balancing* competing considerations. Beauchamp and Childress describe balancing as being of coordinate importance with specification.[33] For Gert, Culver, and Clouser, as we will see, balancing is even more central, on account of their attempt to unify morality around the idea of minimizing harm.

In addition to these schematic similarities, there are ways in which these two views are complementary.[34] This is sometimes obscured by polemical rhetoric in which one of these rivals tasks the other for lacking a certain account. By calling the features I am about to mention "complementarities," though, I am suggesting that there is no reason why these useful aspects of each account could not be adopted by both — or, indeed, by anyone. I have in mind two pairs of such helpful features.

The first pair of complementary contributions arises from the authors' respective attempts to constrain the reasoning whereby prima facie norms get overridden. Gert, Culver, and Clouser offer a helpful checklist of questions to ask in determining whether it is justifiable to violate one of the moral rules.[35] Some of these are quite obvious — for example, what are the harms and benefits? Others are less so. Some, for instance, are culled from other moral theories: What relationships among people are involved? How does the difference between harm foreseen and harm intended as an end or means enter in? Although they cast this list of questions as a list of "the morally relevant features,"[36] this is to oversell it. After all, they themselves go on to utilize a much wider set of morally relevant features, ranging from social expectations of privacy to the proper understanding of death. Still, their list of questions is useful to keep in mind when dealing with conflicts and helpful in explicating what they mean in requiring that all moral agents be able to understand the rationale of any rule violation. Similarly useful is Beauchamp and Childress's partially overlapping list of "conditions that restrict balancing."[37] In particular, they supplement Gert, Culver, and Clouser's question about whether there are morally preferable alternatives, with demands that the infringement be minimized and its negative effects mitigated, also adding that the agent must check the realism of his or her projections of expected harms and benefits.

The second pair of complementary contributions I want to highlight arises in the authors' respective accounts of moral justification. Beauchamp and Childress give a helpful exposition of Rawls's notion of reflective equilibrium,[38] in which justificatory priority is not given in principle to abstract considerations over particular ones or vice versa. As a result, deductive arguments are not privileged over inductive ones. Instead, justification is a matter of making arguments in both directions. Although Gert, Culver, and Clouser give no such exposition of coherence, their methodology is not the "deductivist" one that some have charged it with being.[39] Instead, as I have mentioned, they too view the norms they articulate as having been drawn (inductively) from the common culture. For their part, Beauchamp and Childress lack the stress on public justification that is central in Gert, Culver, and Clouser and that is central to my argument. As the latter authors put it, "To justify morality is to show that morality is the kind of public system that all rational persons would favor as a guide for everyone to follow."[40] It follows, as they indicate, that to be justified in overriding one of the moral rules, one must have grounds that a rational person can impartially and publicly advocate.[41] Putting these two aspects together — reflective equilibrium and publicity — we arrive at Rawls's idea of public justification, in which each rational (and reasonable) individual supports a set of norms in reflective equilibrium, and each publicly accepts as reasonable the basis on which each other person supports them.[42] My sense is that this complementarity of views about justification is not merely a conceptual possibility and that each set of authors would, in fact, endorse their combination. Certainly — and this is important for my later argument — there is no indication that Beauchamp and Childress would dissent from the requirement of publicity stressed by Gert, Culver, and Clouser.

Finally, I come to what is, from my point of view, the salient difference between the two positions. Despite the plurality of rules they put forward, Gert, Culver, and Clouser claim that their account is unified around the idea of preventing harm and pervasively use the idea of minimizing harm in indicating how they would interpret their principles. "The goal of morality," they write, is "to lessen the overall evil or harm in the world."[43] Each of their ten moral rules, they claim, is either a prohibition on harming people in a certain way or else is justified instrumentally on the basis of its importance in preventing harms. Thus, they cast the list of "morally relevant features" as providing instructions for how to balance harms and benefits.[44] Accordingly, while the "relevant features" give shape to their discussions of rule violations, entailing that their view does not coincide with any simple consequentialism, their bottom-line question is nonetheless often put in simple, global balancing terms: Is the harm involved in acting against the rule "greater" than the benefit to be attained by doing so?[45] As they put it,

> the way an impartial rational person decides whether to publicly allow a violation [of one of the ten rules] is by estimating the amount of harm that would result from this kind of violation being publicly allowed versus its not being publicly allowed.[46]

Beauchamp and Childress, by contrast, place no harm-minimization principle at the core, instead resting content with the degree of unification their four principles provide.

"Least Harm" as a Merely Intuitive Standard for Use in Bringing Norms to Bear

I use Gert, Culver, and Clouser's view to exemplify the use of global (type 1b) balancing, as distinguished from conflict-resolving (type 2) balancing. In using the idea of harm-minimization to clothe their view with a semblance of unity, they make global balancing a feature of their theory, one that, they claim, is implied, or at least suggested, by the various moral rules.[47] The question is whether it is plausible to claim that this balancing might proceed in a way that satisfies the requirement of publicity. I argue that it cannot.

It matters greatly, in this respect, whether values are commensurable: whether the reasons against action represented by harms may all be adequately arrayed, for the purposes of deliberation, on a single dimension.[48] On some past theories, which were monistic about value, it would be easy to make sense of the idea of minimizing harm. Harms and benefits all falling on one dimension, all one would need to do would be to compute where on that dimension an alternative fell. Such theories purport to proceed on a publicly assessable basis. Their problem is not with publicity, but with the facts. Gert, Culver, and Clouser, to their credit, do not succumb to such false claims of commensurability. Instead, they insist that there are five qualitatively distinct dimensions of disvalue, five irreducibly different types of bad: death, pain, disability, loss of freedom, and loss of pleasure. That these are qualitatively distinct and not all measurable on one scale seems correct. If this is accepted, though, then their claim that the idea of minimizing harms unifies morality more strongly than do Beauchamp and Childress will be meaningless unless there is a systematic and publicly explicable way to balance these incommensurable harms.

In fact, however, there is none. Gert, Culver, and Clouser certainly offer none. Instead, they fall back on intuitions. Here is one of their examples: Although "wearing an orange necktie with a fuchsia shirt" may displease some people, and one will violate the moral rule against depriving people of pleasure if one wears such a combination with this intention, such sartorial choices are not, in fact, morally prohibited, because the harm involved in depriving people of the freedom to make such choices is "greater" than the harm involved in the displeasure they cause.[49] Now, we must ask, how do Gert, Culver, and Clouser purport to know this? Have they done research about sartorial freedom and color-clash wincing? No; and since harm is not one kind of thing, there is not any obvious research that could possibly settle this question. Instead, it is simply supposed to be intuitively obvious as to which harm is greater. Perhaps we could test the case in thought by varying the hypothetical. What if a speaker wore a brightly colored sixties-era tie with a relatively plain nineties-era shirt? What if the speaker wore nothing at all? Would that cause the audience so much displeasure that it would be morally justified to deprive him or her of this freedom? Would we make moral exceptions for gorgeous people, the sight of whose nakedness, even on a public stage, might result in a net *increase* in pleasure? I submit that the idea of avoiding what results in the greater harm does not name or suggest a usable or publicly explicable way to settle these sorts of issues.

In fact, as these rhetorical questions begin to suggest, the features that affect the balance of resulting pleasure are by no means the most salient among those that matter to the moral appropriateness of one's attire. Given the social conventions that prevail in most of the world, a proper respect for others dictates wearing *some* clothing, irrespective of the possible pleasure produced by going without. Duties of consideration for others interact complexly with evolving traditions to yield more nuanced and specific norms that forbid, say, wearing beach clothes to a funeral.[50] Quantitative balancing of pleasure and displeasure will not track the distinctions that matter in such cases: the pleasure that someone gets from wearing beach clothes to a funeral is rightly discounted. A further fact confirming that the amount of displeasure is not dispositive in these kinds of cases is that the considered judgment with which Gert, Culver, and Clouser begin, which favors individual liberty in the choice of wardrobe within some range, is quite insensitive to the size of the audience. If what mattered were the *amount* of displeasure caused by someone wearing an orange tie with a fuchsia shirt, then we ought to be *adding up* the displeasure of each audience member; and if the audience gets big enough, then it would turn out to be a moral violation to wear such a combination. But that, on due reflection, is neither what we think nor how we think.[51]

In any case, there is no hint of a method here for determining which harms are greater than others. This leaves me with the suspicion that what is actually happening is that Gert, Culver, and Clouser are noticing which moral conclusions we actually come to and then *read off* from those judgments the assertion that we judge one harm to be greater than another. If this is the case — and this is certainly the direction in which economists working with revealed preference theory proceed[52] — then judgments of relative harm can never provide a *way* of determining which norm overrides. Rather, they provide a way of restating the conclusion that one overrides the other.

If we put this together with Gert, Culver, and Clouser's denial that harm is one kind of thing, what we see is that their claim to have unified morality more than have their rivals is spurious. Like their rivals, they identify four or five independent norms. Like their rivals, they see that each of these can sometimes override the others.[53] Although Gert, Culver, and Clouser's push for unification ends up laying more stress on the language of balancing than do Beauchamp and Childress, the former group has no more to offer by way of a weighing procedure. This is ironic, given Gert, Culver, and Clouser's healthy emphasis on the publicity of moral justification. The principal difficulty with the metaphor of balancing, when used to cover a judgment made without any quantitative basis, is that it tends to mask the real reasons at work. At the very least, it fails to encourage the articulation of the real reasons, the kind that a more frankly qualitative account, such as the one I began to sketch regarding nakedness, might bring out. Public justification cannot be built without each person articulating what his or her reasons for judgment are. The metaphor of balancing here provides an excuse for laziness in this regard and does little else. Given the falsity of value commensurability, relying on a principle whose content implies or features global balancing will inevitably clash with the requirement of publicity in these ways: by depending on intuitive quantitative balancings whose basis cannot be publicly expressed because there is no actual quantitative dimension backing them up and by failing to encourage the public articulation of the actual, qualitative bases of such judgments.

Must We Ever Resort to Balancing to Resolve Conflicts among Principles?

The other troubling use of the balancing metaphor is as a way to resolve conflicts among principles. Balancing that resolves conflicts either rests upon articulable reasons or it does not. I will use the term "intuitive balancing" for instances in which the deliberator is unable to articulate his or her reasons for weighing matters differentially. The remaining acts of balancing are based on an underlying reason; these are what Beauchamp and Childress call "justified acts of balancing." These latter, by definition, "entail that good reasons be provided for one's judgment," or at least underlie the judgment implicitly.[54] Accordingly, justified acts of balancing do not directly violate the publicity requirement in the way that intuitive balancing does. Indeed, as defined, justified acts of balancing meet the strong requirement of publicity, which demands that the reasons be actually expressed, not merely publicly expressible. Whereas Gert, Culver, and Clouser are distinctive in the extent to which they rely on (what turns out to be) intuitive balancing, both sets of authors invoke justified balancing of plural considerations as a recommended way of bringing their plural principles to bear on practice. This recommendation I dispute. I argue that to rely on balancing, rather than specification and other modes of interpretation, in dealing with conflicts among principles is to go against the requirements of the cooperative development of action-guiding theory. Instead of balancing norms, we should specify or otherwise interpret them.

One would underestimate the potential for the various modes of interpretation if one thought in terms of a dichotomy between "interpreted" and "uninterpreted" norms. As the history of Talmudic interpretation shows, it is possible to keep adding further layers of interpretation. More generally, it is significant that all four types of interpretation distinguished earlier were defined in terms of a *relation* between two norms: a norm is a specification, or a narrowing, a gloss, or a sharpening, *of another norm.* The ideas of specification and its kin are all *relative to an initial norm.* A norm that is a specification of an initial norm may in its turn be specified. Hence, when Gert, Culver, and Clouser complain of what they term my "failure to formulate any procedure for dealing with conflicts between *specified* norms,"[55] they obscure the point that, just because the norms that are in conflict are specifications of some other norms, this does not mean that the conflicting norms cannot be *further* specified so as to relieve the conflict. Of course, if by "a procedure for dealing with conflicts" they mean a decision procedure that will guarantee more automatic results than can the process of specification, which rests on deliberative rationality, then I do not believe in any such procedure, nor have they offered one. What allows the idea of specification to offer a third way of reflectively coping with conflicts among principles is the fact that it offers a change in the set of norms. It will be important to keep this possibility of continuous or progressive specification in mind as I turn to the question of whether justified balancing is needed as a distinct mode of addressing conflicts among principles.

To be sure, not all conflicts among moral principles are resolvable. This is something I explicitly noted in my original article on specification. Contrary to what Gert, Culver, and Clouser suggest,[56] I did not there "fail to realize that some moral disagreements are not resolvable" — a failure, they think, that explains the purported further failure (dissolved in the last paragraph) to provide any procedure for dealing with conflicts among "specified" norms. Setting aside their conflation of conflicts and disagreements, I note that I explicitly allowed for the possibility of genuine, unresolvable moral dilemmas, that is, for unresolvable conflicts of the strongest sort.[57]

As I have noted, Beauchamp and Childress are quite supportive of the idea of progressive specification. They insist, however, that justified balancing ought to be retained as a distinct and complementary mode of dealing with conflicts among principles. Against the suggestion of simply submerging the idea of balancing into that of specifying, they respond as follows:

> Balancing often eventuates in specification, but it need not; and specification often involves balancing, but it might only add details. Accordingly, we do not propose to merge the two methods. The point is that balancing does not compete with specification, and they both coherently augment the model of coherence. We therefore propose that balancing and specification be seamlessly united with a general model of coherence that requires us to defend the reasons we give for actions and norms. . . . [B]alancing is particularly useful for case analysis, and specification for policy development.[58]

In the body of their book, they follow through on this stance, calling in many places for "further specifying and balancing" of principles.[59] The narrow question on which I part company with them is whether balancing is ever to be recommended as a distinct mode of resolving a conflict among principles. In effect, my argument will build on the observation that, in a context such as bioethics, at least, case analysis cannot be separated from policy development.

One way to restate this question about this conflict-resolving role of justified balancing — intuitive balancing having been disposed of by my argument in the last section — follows: given that one has a reason for resolving a conflict one way rather than another, what compelling reason might one have for refusing to incorporate that reason into further specification of one or the other of the competing principles?[60] This, in effect, is what we saw happen in the sartorial cases of the last section. There, we ended up with something like "one's freedom to wear what one wants in public, despite offense one may give to others, is to be respected so long as one at least wears something." While Beauchamp and Childress are correct in noting that justified balancing does not always eventuate in specification, it seems that it always *can*. Consider the following simple example, which is one of the rare instances in which Beauchamp and Childress mention conflict-resolving balancing without linking it to specification: the case of a conflict between the virtues of generosity and tolerance and a duty (perhaps of justice) that calls for outrage or punishment.[61] Now, suppose that the reason or consideration that justifies tipping the balance in favor of punishment is as follows: "While the virtues of generosity and toleration call for us to be generous to all *persons*, some *behaviors* are beyond the pale and demand that we express our outrage and punish them severely." Plainly, however, this reason embeds a distinction that will be incorporated into our interpretation of these virtues. It can generate the following specifying move: from "be generous and tolerant" to "be generous and tolerant toward all persons even when they have transgressed, but toward their behavior only when that behavior is within the pale." This specification, in turn, could well guide a wiser policy of treating transgressors as not irredeemably bad.

Let me give another example where specification is a more fruitful and explicit way of resolving a concrete issue than is balancing. Again, I am not a bioethicist and do not purport to judge the soundness of the reasoning I will describe in this paragraph. Instead, I offer as an example a speculative reconstruction of some actual reasoning that seems sensible, even if its results are not correct. What matters here are the kinds of moves that specification makes available. Historically, Beauchamp and Childress's principles arose in tandem with the *Belmont Report*'s principles governing research ethics. One issue that has continued to be hotly debated is how to treat research that is both carried out on and intended to benefit children. Children cannot meaningfully give consent, yet sometimes the potential benefits of a proposed research protocol are so great that it seems crazy to block it solely on that basis. Because the *Belmont* principles are so highly entrenched in this area, it sometimes seems to the commentators writing on these issues that the question is whether the principle of autonomy should be "balanced" against the principle of beneficence (or perhaps justice, if the relevant group of beneficiary children is not well off) or whether, instead, the principles should be "ranked" against each other in some lexical fashion that prohibits these trade-offs within certain ranges.[62] To approach these questions using this dichotomy, however, is to fail to make explicit the normative stance that pervades this whole context, namely, a presumption against the permissibility of research on human subjects.

To make this normative stance explicit, we should recognize that, in addition to the general principles of autonomy, beneficence/nonmaleficence, and justice, we also (if we accept the content of *Belmont*, or something approximating it) start out with a principle that unites them and specifies how they are to be brought to bear on the research context. It has the form, "It is impermissible to engage in research on human subjects unless the principles of autonomy, beneficence, and justice are adequately satisfied." Call this the "protean research-limiting principle."

With reference to this protean principle, we may fruitfully recast the debate about whether the general principles of autonomy, beneficence, and justice were to be "ranked" or "balanced." The idea of specification helps us articulate the different interpretive alternatives, which show up in two possible ways of specifying the protean principle's vague notion of "adequate satisfaction." The less restrictive specification is "It is impermissible to engage in research on human subjects unless the principles of autonomy, beneficence, and justice are *satisfied on balance*." The more restrictive specification is the following: "It is impermissible to engage in research on human subjects unless *we do so in a way that respects their autonomy, proceeds justly, does no (intentional?) harm, and produces (significant) benefits*." Call this "the restrictive research-limiting principle."

Once the research-limiting principle is spelled out, however, we can see that the debate about whether to allow trade-offs among autonomy and beneficence in the case of research on children can be recast as a debate about whether to qualify or specify this one principle in a way that is relevant to the differences between children and adults. The National Commission apparently did make a distinction between children and adults, a distinction that showed up in the different ways they specified the restrictive research-limiting principle for the two cases. In neither case did they intend the principle to be absolute or to be applied without qualification. One qualification had to do with degree of risk. In the case of adults, this specificatory qualification, in effect, read as follows: "It is impermissible to engage in research *posing more than minimal risk* to human subjects unless. . . ." Given the inability of children to give meaningful consent, however, the restrictive research-limiting principle, so specified, still conflicted sharply with the aim of benefiting children. Accordingly, the National Commission put forward a tentative compromise that in effect attempted to resolve this residual conflict by specifying the restrictive research-limiting principle differently in cases involving children. What they suggested was a principle beginning as follows: "It is impermissible to engage in research posing *more than a minor increase over minimal risks* to human subjects *who are children* unless. . . ." The clause pertaining to expected benefit was compensatingly beefed up.

As is plain from this example, the bare idea of specification does not indicate how one ought to specify principles so as to resolve a concrete problem, nor is it meant to. Multiple alternative, incompatible resolutions might be reached by specifying (as by intuitively balancing). Being discursively explicit, however, specifications can be defended on the basis of reflective equilibrium: by making arguments that show how they may be supported by their fit with what we continue to believe, on due reflection. Having a substantive ethical theory that one is working out will be practically indispensable in making strong connections of this kind and will obviously help narrow down among possible specifications. Conversely, making explicit connections by specifying norms is indispensable to making progress in ethical theory. The kind of chain of specification exemplified by my reconstructed National Commission example enables one to connect conclusions to initial principles while at the same time developing more nuanced and definite guidance that responds to what are taken to be morally relevant distinctions among different concrete situations.

Can we imagine any reason that justifies a balancing outcome that *could not* be incorporated into a specification in this way? One apparent possibility is that some third norm might come in as a contingent tiebreaker. Suppose we are undecided as to whether it is permissible for the pauper to steal medicine from the pharmacist in order to save his wife's life. If so, then the fact that the pharmacist happens to have promised his entire supply of that medicine to someone else may tip the balance. Further, because the connection between stealing and promise breaking is, here, entirely contingent, it seems hasty to try to build it into some further specification of our norms. But this only appears to be a case of justified balancing. It obtains that appearance from the speciously precise suggestion that one began with some sort of quantitative tie. Once that false suggestion is subtracted, it is no longer apparent that two principles win out over one. To settle that question, one is thrown back on intuitive balancing, after all. To be sure, if the compelling aspect of the additional consideration is not the promise, per se, but the fact that someone else needs this life-saving medication, then a possible specification is in the offing: "One may not steal medicine to save the life of someone one loves when doing so deprives someone else, legally entitled to them, of the life-saving properties of that medicine." I cannot imagine a case of justified balancing that could not generate a specification.

If we can always specify in cases in which we have justifying reasons for resolving a conflict among bioethical principles, then we should — with only one caveat, to be mentioned shortly. To explain why we should, I revert to the second and third assumptions I stated at the outset of the chapter: that bioethics aims to guide action and that its doing so requires the progressive collaboration of many practitioners and theorists. Now, like both sets of authors I have been discussing, I view the progress of ethical thinking, even when it goes well, as halting and tentative. It is not a matter of monotonically filling out a sketch. There will be erasures. Whole patches will have to be painted over and begun afresh. An important part of the progress, however, will be improvement in sorting out conflicts in advance. To the extent that this is not done, the theory is not guiding action clearly. Indispensable to the progress of the enterprise in sorting out conflicts is the public airing of hypotheses: of candidate specifications and interpretations that are put forward for consideration as ways of resolving like conflicts in the future. In order to contribute to this broader enterprise, then, theorists and deliberators who have the opportunity to articulate the resolution of a conflict in the form of a specified principle ought to do so. Only by doing this will they project their resolution to future cases. Therefore, specification, which seems always to be possible in cases in which balancing is justified, always supersedes balancing.

The one caveat I want to enter is that sometimes we may feel so tentative about the resolution we have reached in a concrete case that we feel we are not in a position to project it into the future. Sometimes we will do better to admit that a problem needs further work. Notice, though, that in such instances, resting with balancing is being commended only weakly, as a second-best outcome.

Is There Any Place for Balancing in Bioethics?

Do I, then, recognize no appropriate role for balancing, apart from the caveat just entered? The one type of balancing distinguished earlier that I have not yet addressed is contextual or piecemeal balancing that enters as a feature or implication of the content of an ethical principle. This limited sort of balancing must be allowed.

A good example of this arises within Beauchamp and Childress's discussion of the principle of nonmaleficence. Essential to medical contexts is the need to interpret this principle in light of the idea of net harm or benefit or in terms of the balance of harms and benefits. Now consider what obligations *not* to treat might arise from this idea. In the case of competent patients, Beauchamp and Childress suggest, there is no need for the relevant principle to call for us, the moral assessors, to balance the harms and benefits. Instead, emphasis falls on patient consent, the patient being deemed well able to balance the harms and benefits as he or she sees fit. In the case of incompetent patients, however, there seems to be no good alternative to having the moral assessors do their own balancing. By this route, we arrive at a conclusion, such as "the burdens [of treatment] can so outweigh the benefits to the incompetent patient that the treatment is wrong rather than optional."[63] Here we have a principle that does not put forward a global balancing principle; rather, it confines balancing to a special case within the adaptation of the principle of nonmaleficence in the context of medical treatment. Local, context-specific (type 1a) balancing of this kind seems a sound feature of our principles, and one that is difficult to eliminate at this level of abstraction. (It may turn out, however, that *further* specification of this principle will move us away from the appearance of seeking a quantitative "netting out" of burdens and benefits and toward a far more context-sensitive and precedent-informed set of judgments about which burdens "so outweigh" prospective benefits that treatment ought not to be pursued.)

I conclude, then, not that there is no appropriate place for balancing in bioethics, but that its place is limited to contexts that are both relatively narrow and shaped by surrounding principles. Global (type 1b) balancing, attractive to some as a way of seemingly bringing unity to ethical theory, is ruled out by the principle of publicity, which is especially important in a domain of public-policy ethics such as bioethics, in conjunction with the fact of value incommensurability. Conflict-resolving (type 2) balancing, whether intuitive or "justified," is ruled out by the superiority of specifying norms (and otherwise interpreting them) in such contexts. This superiority, as I have argued, lies in the greater contribution of specifying and interpreting norms to the overall enterprise of progressively developing action-guiding principles. Contributing to this enterprise is part of the purpose of work in bioethics.

Notes

The original version of this chapter was presented at a conference commemorating the twentieth anniversary of the *Belmont Report*, held at the University of Virginia in April 1999, and a subsequent version was published in *Journal of Medicine and Philosophy* 25 (2000): 285–307. I am grateful to James F. Childress and Harold Shapiro for their initial invitation to participate. I have benefited greatly from several rounds of useful comments from Tom L. Beauchamp and from the criticisms of James F. Childress, Ezekiel J. Emanuel, Jorge L. A. Garcia, and Robert M. Veatch. I also make some effort here to clarify my disagreements with Bernard Gert, Charles Culver, and K. Danner Clouser, drawing on their "Common Morality versus Specified Principlism: Reply to Richardson," in the same issue of *JMP* at 308–32.

1. Henry S. Richardson, "Specifying Norms as a Way to Resolve Concrete Ethical Problems," *Philosophy and Public Affairs* 19 (1990): 279–310.

2. I am, in fact, quite indebted to the writers I will be discussing for having raised many important questions about the idea of specification. In addition to Beauchamp and Childress (*Principles of Biomedical Ethics*, 4th ed. [New York: Oxford University Press, 1994, hereafter B&C]), special mention is due to David DeGrazia, "Moving Forward in Bioethical Theory: Theories, Cases, and Specified Principlism," *Journal of Medicine and Philosophy* 17 (1992): 511–39 — though I should say that I do not count myself as a proponent of "specified principlism." I have benefited from many instructive discussions with Dr. DeGrazia over the years.

3. I find Gert, Culver, and Clouser to be somewhat disingenuous about this

in "Common Morality versus Specified Principlism," 316. There, they write that "we do not attempt to generate answers from our moral theory, but rather to describe the answers that are given by common morality, and to determine whether those answers can be justified." Supposing that they mean that their theory says that, indeed, everything that common morality holds (and only that) is well justified, even this would be action guiding, for it would remove doubts. Further, the set of all individuals is not of one mind about all the provisions of common morality, and such a justification, insofar as it became accessible, would help bolster a fuller convergence of belief. In any case, it would woefully understate the amount of productive work Gert, Culver, and Clouser have done to say that they leave the content of common morality just as they found it.

4. Here I endeavor to be as clear as possible that the reliance on balancing I see in Gert, Culver, and Clouser's account does not constitute, and is not supposed by me to constitute, a mode of conflict resolution. Although the earlier version of this chapter in the *Journal of Medicine and Philosophy* similarly described their view as one relying on "global balancing" (286), in contrast to Beauchamp and Childress's view, which relies mainly on conflict-resolving balancing (I wanted one example of each type), much of Gert, Culver, and Clouser's reply to me in "Common Morality versus Specified Principlism" misunderstands this and proceeds as if I had claimed that they heavily rely on balancing to resolve conflicts. See especially their p. 315, where they quote my characterization of Beauchamp and Childress as if I had applied it to them.

5. Bernard Gert, Charles Culver, and K. Danner Clouser in *Bioethics: A Return to Fundamentals* (New York: Oxford University Press, 1997) [hereafter GCC].

6. In the time since this article first appeared, Beauchamp and Childress have brought out a new edition of their textbook: Tom L. Beauchamp and James F. Childress, *Principles of Biomedical Ethics*, 5th ed. (New York: Oxford University Press, 2001). Although I have not been through this new edition with any thoroughness, I note some new points about the limitations of specification at p. 19:

> It also seems pointless or unduly complicated to engage in specification in many circumstances. For example, in individual cases of balancing harms of treatment against the benefits of treatment of incompetent patients, the cases are often unique or so exceptional that it is perilous to generalize a conclusion.

There are several distinct points to note here. First, there is the question of who is offering a suggested resolution of the unique, individual case. If it is an ethics consultant, it seems to me that the consultant owes the patient or the patient's family something more articulate than "this is how the balance seems to me to fall out in this case." Whether the patient him- or herself owes anyone a more articulate reason will depend on the context. Second, there is the possibility that the balancing that Beauchamp and Childress are here considering is explicitly *called for* by the relevant norms pertaining to the treatment of incompetent patients. If so, this would be an instance of the kind of "contextual balancing" that, in the final section of this chapter, I accept as fully legitimate. Finally, I readily agree that there are always perils in generalizing from any case or set of cases.

7. See Henry S. Richardson, "Institutionally Divided Moral Responsibility," in *Social Philosophy and Policy* 16 (1999): 218–49; and Ellen Frankel Paul et al., eds., *Responsibility* (Cambridge: Cambridge University Press, 1999): 218–49.

8. Unless a norm is universally quantified, in the logicians' sense (beginning "for all . . ."), deductive subsumption cannot work with them. Contrary to GCC, 89, I did not confuse and am not confusing the absoluteness of logical form, which logicians designate a "universal generalization" (as opposed to an existentially quantified statement), with universality or universalizability across the domain of persons or moral agents.

9. Accordingly, in defending the fruitfulness of specifying, it is not part of my task to develop substantive criteria for moral decision making. Gert, Culver, and Clouser write that I "cannot" develop such criteria (GCC, 89), and I certainly do not claim that the idea of specification can generate such criteria from thin air. Instead, my question is whether, in bringing to bear morally relevant considerations on practical problems, specification or balancing offers a more productive model of how to proceed.

10. In "Beyond Good and Right: Toward a Constructive Ethical Pragmatism," *Philosophy & Public Affairs* 24 (1995): 108–41, I argue that the idea of specification suggests how to steer between the rocks of consequentialism and deontology, but I do not presume the correctness of this argument here.

11. Robert M. Veatch, "Resolving Conflicts among Principles: Ranking, Balancing, and Specifying," *Kennedy Institute of Ethics Journal* 5 (1995): 210, 216.

12. It is possible that, rather than ignoring the possibility of specifications that distinguish between different contexts, Veatch is implicitly asserting that we do not need to make distinctions among contexts in stating correct moral principles. To the contrary, I believe that we do. For instance, I believe that in *A Theory of Justice* (Cambridge, MA: Harvard University Press, 1971), John Rawls made real moral progress in grappling with the conflict between the vague values of freedom and equality by developing a specification of them that reconciled their demands in a way that was tailored to a specific context, namely, the appropriate configuration of "the basic structure of society."

13. Cf. the characterization of the National Commission's work in GCC, 73. In reality, of course, specificatory and abstracting work need to be interleaved.

14. GCC, 53.

15. This divergence between interpretation and derivation is related to the logical form of the initial norm. If it is a universal generalization, all instances may be derived from it by deductive subsumption. If it is not, however, interpretive glosses will be needed to indicate which of its potential implications are to be counted as following through on the norm. In this way, interpretation modifies norms that are logically loose enough to permit this sort of supplementation. As I argued in "Specifying Norms," most of our norms are loose enough for this.

16. Gert, Culver, and Clouser claim that I do "not recognize that in order

to make norms culturally sensitive, it is necessary to allow for some degree of interpretation of the norms" (GCC, 89). Now, specification being a mode of interpretation, it is plain that I allow for some degree of interpretation. Although my 1990 article did not discuss cultural variations or how to account for them, in "Aristotelian Social Democracy," in *Liberalism and the Good*, ed. R. Bruce Douglass et al. (New York: Routledge, 1990): 203–52, 235, Martha C. Nussbaum develops the idea of "local specification" to do just that.

17. Gert, Culver, and Clouser accuse me of failing to recognize that norms that are not logically absolute, and hence allow exceptions, can nonetheless be universal in the sense that "they apply to all rational persons" (GCC, 89). This accusation is a mistake, which derives from their not noticing in the relevant passage ("Specifying Norms," 292–95) that the only sort of universality I mention is not universality over the domain of persons but universality over the domain of acts ("for all acts . . .").

18. B&C, 39.

19. We might push for analysis to firm up our norms sufficiently to allow specification to begin. What "analysis" should mean, in this context, is a deep question. For arguments that autonomy cannot be given an analysis in terms of necessary and sufficient conditions, see Gerald Dworkin, *The Theory and Practice of Autonomy* (Cambridge: Cambridge University Press, 1988), esp. 6, and my "Autonomy's Many Normative Presuppositions," *American Philosophical Quarterly* 38 (2001): 287–30.

20. To solidify this implication, I should have spelled out that the only syntactic changes to the norm were those that flow from glossing the determinables.

21. For additional reasons to regard the two views as complementary, see Tom L. Beauchamp, "Principlism and Its Alleged Competitors," *Kennedy Institute of Ethics Journal* 5 (1995): 181–98.

22. B&C, 6, 37, 101; GCC, 16–17, 33–34.

23. GCC, 34.

24. B&C, 33–36.

25. GCC, 37. For more detail, see Bernard Gert, *Morality: Its Nature and Justification* (New York: Oxford University Press, 1998), chap. 9.

26. GCC, 26, 28.

27. GCC, 31. See also Gert, *Morality*, chap. 6.

28. As I will suggest in the following section, some of the moves that Beauchamp and Childress describe as specifications I would count as interpretations of a more generic kind. Gert, Culver, and Clouser discuss interpretation of the rules in GCC, 54–60.

29. B&C, 28–32; also see index *s.v.* "specification."

30. GCC, 91.

31. GCC, 52–53.

32. GCC, 55.

33. B&C, 32–34.

34. Beauchamp notes the fact in "Principlism and Its Alleged Competitors."

35. GCC, 38.

36. GCC, 37.

37. B&C, 34.

38. B&C, 20–28.

39. Beauchamp and Childress acknowledge this in B&C, 20. Gert and Clouser are charged with "deductivism" by, e.g., B. Andrew Lustig, "The Method of Principlism: A Critique of the Critique," *Journal of Medicine and Philosophy* 17 (1992): 487–510.

40. GCC, 26.

41. GCC, 37.

42. John Rawls, *Political Liberalism*, rev. ed. (New York: Columbia University Press, 1996), 387.

43. GCC, 62. Cf. GCC, 80, on moral rules and moral ideals. That their theory is unified around the idea of minimizing harm is the only apparent basis for Gert, Culver, and Clouser's claim (GCC, 20, 88) that they recognize the unity of morality in a way that Beauchamp and Childress do not.

44. GCC, 86.

45. E.g., GCC, 56, 245. At GCC, 58, Gert, Culver, and Clouser indicate that the calculation of benefits and harms is to be supplemented by a potentially qualitative public consideration of the acceptability of a proposed modification to the rules. In this chapter, I am in effect arguing that the latter standard of acceptability on due reflection, or of reflective equilibrium, should be taken to supersede any purported notion of global, quantitative balancing.

46. Gert, Culver, and Clouser, "Common Morality versus Specified Principlism," 318. Thus, in claiming that they rely on global (type 1b) balancing, I do indeed have in mind "the second step of the two-step procedure that an impartial person [according to them] uses when deciding whether to publicly allow a violation." Ibid., 312.

47. For present purposes, I take at face value Gert, Culver, and Clouser's claim to have unified their theory in this way. For doubts as to whether they actually succeed in this, see my "Bioethics: Root and Branch," *The Hastings Center Report* (September–October 1999): 40–42.

48. See the definition of "value commensurability" in Henry S. Richardson, *Practical Reasoning about Final Ends* (Cambridge: Cambridge University Press, 1994), sec. 16.

49. GCC, 56.

50. Here I concur with Gert, Culver, and Clouser about the interaction of morality and convention. Cf. GCC, 53, with my "Institutionally Divided Moral Responsibility."

51. For a nuanced discussion of the limitations on aggregative thinking, see T. M. Scanlon, *What We Owe to Each Other* (Cambridge, MA: Harvard University Press, 1998), chap. 5. For elaboration of the suggestion that moral theory should draw on modes of deliberation that we accept on reflection, as well as on judgments that we accept on reflection, see the outset of my "Beyond Good and Right."

52. See my criticism of preference-based models of deliberation in *Practical Reasoning about Final Ends*, sec. 15.

53. Gert, Culver, and Clouser at one point even say that moral ideals can sometimes override the moral rules (GCC, 21) — and this despite their heavy criticism of Beauchamp and Childress for failing to take seriously enough the distinction between the two categories of norms (B&C, 77).

54. B&C, 33.

55. GCC, 89, emphasis added.

56. GCC, 89.

57. Richardson, "Specifying Norms," text at n. 48.

58. B&C, 34.

59. B&C, 37, 101–2, 107, 328, 331, 334, 412, 433, 471.

60. Beauchamp and Childress mention David DeGrazia as having noted the possibility of using the reason that makes the balancing "justified" to generate a specification in this way. See B&C, 34.

61. B&C, 67.

62. I am here simplifying a more subtle discussion in Robert M. Veatch, "Ranking, Balancing, or Simultaneity: Resolving Conflicts among the *Belmont* Principles," in this volume. I also draw from this chapter the factual basis (such as it is) for my speculative reconstruction of the National Commission's reasoning in the following text.

63. B&C, 212.

15 Max Weber Meets the *Belmont Report*

Toward a Sociological Interpretation of Principlism

John H. Evans

It is well accepted that the bioethical principles first articulated in the *Belmont Report* and in influential textbooks[1] have in many ways structured decision making in biomedical ethics.[2] These principles have not just had an impact on the interaction between physician and patient, but on health policy more broadly, serving as the ethical standard on health care policies that range from whether the government should allow and pay for human genetic engineering to President Clinton's health care reform plan of the early 1990s. In this chapter I do not address whether the current principles are correct or apply them to problems. Rather, I intend to explain (1) why the principles are used in decision making, (2) the power of some of the critiques of principles, (3) the future prospects of those critiques, and (4) the costs and benefits we as a society incur for the use of institutionalized principles in public ethical decision making. It is generally agreed that, to understand the use of principles, often referred to as principlism, we must go back in history, perhaps to the Nuremberg trials. I propose that to understand principlism we must go much further back in time.

In 1494 Frater Lucas Pacioli wrote what is widely recognized as the first and most influential textbook for double-entry bookkeeping. This is the tabulation that any academic department chair is familiar with: what are our costs for the next year, and what is our income? And, more specifically, are the costs associated with this component of the business generating returns that justify the costs? Max Weber described the innovation in this form of accounting as allowing for

> the valuation and verification of opportunities for profit and the success of profit-making activity by means of a valuation of the total assets (goods and money) of the enterprise at the beginning of the profit-making venture, and the comparison of this with a similar valuation of the assets still present and newly acquired, at the end of the process.[3]

This process is so taken for granted it is hard to imagine an alternative. Before the invention of this system, however, accounts from businesses were basically a "rambling story with numbers" that served to "assist the memory of the businessman" but not help with evaluation of the businessman's actions.[4] Double-entry bookkeeping was thus a major innovation in economic history. A couple of changes in the accounts system also transformed it into a formally rational procedure that allowed for calculability, efficiency, and predictability in human action. The first change was that the new system became a means of throwing away information deemed to be extraneous to the decision-making process. The second change was that these numbers took on a new degree of calculability. Instead of there being proceeds on one list and costs on another, these two were *translated* into a common metric called "profit," which made an evaluation of each action much more precise. With the previous "rambling narrative" style of bookkeeping, owners could not readily determine whether an action (such as delivering to the abbot) was efficient at maximizing their end (profit). With the formally rational accounting that combined information about the costs and proceeds, the efficacy of a sale could be calculated.

The relevance of fifteenth-century bookkeepers to our current bioethical enterprise is that Weber thought that early formally rational institutions such as double-entry bookkeeping would spread and eventually displace practices based on other forms of rationality. Weber believed that what he called the "iron cage" of formal rationality — represented in the new accounting and other practices — would continue to tighten, "perhaps until the last ton of fossilized coal is burnt."[5] Or, in the words of a Weber interpreter, Weber believed that a formally rational social life "is no mere possibility, but the inescapable fate of the modern world."[6] To continue, and possibly to mangle the metaphor beyond all usefulness, in recent years we have seen the "iron cage" encircle ethical decision making in medicine and science in the form of the creation, spread, and institutionalization of the formally rational *Belmont* principles.

The *Belmont* Principles as a Formally Rational Institution

Was there a pre-formally rational method of decision making in ethics, equivalent to the "rambling narrative" style of bookkeeping? K. Danner Clouser describes medical ethics in the 1960s as "a mixture of religion, whimsy, exhortation, legal precedents, various traditions, philosophies of life, miscellaneous moral rules, and epithets."[7] With a system such as this, how an ethical decision would be made was not calculable and predictable to anyone but the person making the decision, a point to which I will return later.

The *Belmont* principles do in fact offer the lure of calculability and predictability not offered by the jumble described by Clouser. This calculability and predictability — this formal rationality — can be summarized by the notion of commensuration. Commensuration is a method of "measuring different properties normally represented by different units with a single, common standard or unit."[8] Philosophers will immediately recognize utility as one such commensurable metric. Scientists will recognize risk-benefit analysis, which translates all of the information of a situation into a universal commensurable metric of pleasure/pain, allowing for ease of decision making. And most of us recognize money, the most common commensurable metric of all.[9]

As decision-scientists have repeatedly observed, due to cognitive limitations people do not consider all of the possible information in a given problem, but must make decisions by throwing away some relevant information.[10] It is also recognized that systems of commensuration are methods for throwing away information in order to make decision making easier by ignoring aspects of the problem that cannot be translated to the common metric.[11] If, for example, we consider the complexity of a person's values to be reducible to utility, as economists do, then all of the information about the particularity of those values can safely be discarded.

The *Belmont* principles are a form of commensuration, although not as pure a form as money or utility, and not as commensurable as some critics would like (see the following discussion). If money or utility represents one commensurable scale, the principles represent three metrics, with no agreed-upon method of deciding what to do when the metrics on the three scales point in different directions.

Yet the principles are a system of commensuration nonetheless. They are a method of taking the complexity of actually lived moral life and translating this information into three scales by throwing away information that resists translation. To see the principles as commensuration, we can ask the following: Why are there three principles in the *Belmont Report* and not ten, or perhaps twenty principles? Why did a member of the National Commission complain that there were "too many principles" in an early draft of the *Belmont Report* (there were seven at the time) and that the list was not "crisp enough?"[12] The principles were created to enhance calculability or, in more common language, to simplify bioethical decision making. For example, Beauchamp says that principles "provide frameworks of general guidelines that condensed morality to its central elements and gave people from diverse fields an easily grasped set of moral standards."[13] This calculability or simplicity is gained by throwing away information about deeper epistemological or theoretical commitments. With Beauchamp, a professed "rule-utilitarian," and Childress, a professed "rule deontologist," this common metric of principles allows for ethical decision making they can both agree to, despite the massive amount of information about their ethical inclinations represented by the phrases "rule-utilitarian" and "rule-deontologist."[14] They find that "many forms of rule utilitarianism and rule deontology lead to identical rules and actions," that is, there is a commensurable metric between the two.[15] Yet "rule-utilitarian" must provide some information about how to make a decision or the authors would no longer need such labels for themselves. Decision making, despite long-standing differences between utilitarians and deontologists (not to mention theological groundings of various stripes), can be more efficiently calculated with the principles, which Jonsen calls "the common coin of moral discourse."[16]

What does this retelling of the bioethicists' craft from a sociological perspective get us? There are four tasks that can now be better undertaken with this new lens on the principles: (1) an explanation for why the principles were created and why they rose to dominance in the field, (2) an explanation of the attractiveness of a number of critiques of principlism and predictions about the outcomes of these debates, (3) an outline of the benefits and costs of using principles in our moral life, and (4) for those who want to minimize the costs of using principles, suggestions for improvement.

Why Were the Principles Created? Why Have They Prospered?

I will start with an observation we all know, but which serves as a useful reminder. The principles are not designed to assist in individual level decision-making, although they may be insightful in that regard. Rather, they are designed to help people make only those moral decisions that they must later legitimate to a larger group. That is, the principles are not useful for my personal decisions, but rather are useful for making policy-type decisions in the government, in bureaucratic hospitals, and in scientific institutions.[17] People in these institutions find a commensurable set of ethical decision-making principles to be useful compared with alternative methods of decision making, and it is this usefulness that explains the rise and success of principlism. I will examine the fit between principlism and the institutions in turn.

THE STATE

It is well known that the *Belmont* principles were created at the urging of the state and enacted as regulations, yet this fact is rarely applied to our understanding of these principles. The original congressional mandate to the National Commission included the requirement to "identify the ethical principles which should underlie the conduct of biomedical and behavioral research with human subjects and develop guidelines that should be followed in such research."[18] Moreover, after the submission of the *Belmont Report* to the Department of Health, Education and Welfare, the principles eventually "became public law governing the research activities of federally funded scientists."[19] Jonsen later concluded that the principles had "met the need of public-policy makers for a clear and simple statement of the ethical basis for regulation of research."[20]

We must step back and ask why policy makers needed a "clear and simple statement of the ethical basis for research," that is, a commensurable ethical metric that allows the throwing away of information to reduce the complexity of a decision. Put differently, why wasn't the state satisfied with the full complexity of ethical decisions as it had been practiced before this point?

Put simply, it is part of the U.S. political culture not to trust authority, especially government authority, and the authority of bureaucrats in particular. A complex decision means that we would have to trust the judgment of the government functionary, because they cannot readily explain how they reach their decisions. In liberal democracies, decision making must be transparent. Unlike decrees coming from the subjective perspective of the European sovereign, the U.S. political system was partly founded "on the idea that politics is transparent, that political agents, political actions, and political power can be viewed."[21]

Moreover, "in a country where mistrust of government is rife, the temptation to substitute supposedly impersonal calculation for personal, responsible decisions . . . cannot but be exceedingly strong."[22] Note that in other countries government officials are, as Theodore Porter observes, "trusted to exercise judgment wisely and fairly. In the United States, they are expected to follow rules."[23] This desire for the rules of "impersonal calculation" using commensurable metrics has been applied throughout government, as Porter aptly describes, particularly metrics such as cost-benefit analysis.

It has been noted that if Americans do not trust the government making decisions for them, they especially do not trust the government making what are construed to be moral decisions. Therefore, an Institutional Review Board (IRB) that is indirectly representing the government cannot simply approve research into human genetic engineering (HGE) because they think that it is a good idea, but rather must "show" or, better yet, "prove" with scientific methods, that HGE is good for society.[24] They must show their reasoning in a manner that the public can judge. "Weighing" risks and benefits — notice the simple decision-making rule — purports to be more transparent. Like the allure of cost-benefit analysis, the public can feel that it understands the decision being made on its behalf, giving the decision legitimacy.

THE BUREAUCRATIC HOSPITAL AND INSURANCE COMPANIES

The principles have also become influential due to the need to educate physicians, either during medical school or after. Once again, it is not simply their usefulness "by the bedside" that accounts for their popularity, but because they can readily be thought of as a reduced metric for decision making, akin to the rules preferred by bureaucratic institutions. As David Rothman notes, the era of the rise of bioethics and the *Belmont* principles was also the era of the decline of trust in physicians.[25] If we do not know our physicians — due to the rise of managed care, group practices, as well as increasing residential mobility — we do not know their values. At this point, we turn to observable rules such as decision-making metrics in order to create trust.

Rules and procedures are an inescapable component of bureaucracies, and the bureaucratization of medical decision making is on the rise. With conflicts among physicians, patients, and hospitals quickly turning to law for resolution, there is an increasing desire for observable, explicit rules for decision making. Moreover, as managed care creates another level of bureaucracy, often physically quite distant from the physician and the patient, even more rules are necessary. The principles serve as rules, an institutionalized method that, while imperfect, is an "improvement" over alternatives.

SCIENCE AS A DOMINANT INSTITUTION

In my research on government ethics committees I find that there is a tendency for scientists to find theologians unintelligible. Although I have not studied the case in depth, when I attended the National Bioethics Advisory Commission (NBAC) meetings on cloning, I sensed that when evangelical Protestant theologian Gilbert Meilaender spoke, a thick haze rose from the table, levitating perfectly equidistant between Meilaender and the scientists. The common way of explaining this often-noted phenomenon is that theologians use a language all of their own.[26]

The common metric of the principles offers, I believe, a language that scientists can understand. Reductionism is, after all, one of the hallmarks of modern scientific thinking. Consider the observation, made by medical sociologists, that students learn in medical school to objectify the parts of their patients' bodies.[27] In my terms, they are throwing away information about the rest of the body — not to mention the mind, the soul, and the feelings of the patient — in order to make medical decision making possible. Considering all aspects of the patient would make a decision too complex. Reducing ethical elements to a common metric, in this case a set of three in the *Belmont Report* (four in Beauchamp and Childress's textbook), is likely an appealing way of grappling with a question, in much the same way as considering the patient's heart as being separate from the patient's emotions.[28]

Consider why a scientist would find the testimonies of Gilbert Meilaender and of John Robertson on the same issue to be disconcerting. Both are making claims to moral authority, but are using different paradigms or different epistemological references. Both cannot be correct! To the single paradigm worldview so defended by scientists, there may be something deeply flawed about a decision-making process where participants cannot agree on the grounds by which they evaluate correct and incorrect arguments. I suspect that the creation of a commensurable metric for making decisions into which Meilaender and Robertson must translate their claims gives scientists a sense that there is indeed one decision-making strategy in use.

As others have noted, ethicists have no political constituency[29] and in realpolitik terms can rely only on moral suasion. Despite the gains in authority that bioethicists have made in the past few decades, when push comes to shove, ethicists must negotiate with scientists who may have different interests. In my analysis of the HGE debate, I find throughout the 1980s and 1990s that there are some people writing about the ethics of HGE using the principles and some who do not. It is, however, only the principlists whose texts become influential in the debate, who take on positions of authority in decision making, or who directly or indirectly work for scientific institutions. Is this because their work is "better"? Work is only better in reference to institutionalized standards, and scientists have in many ways set those standards. I conclude that principlism was selected because it articulated with the formal rationality embedded in the scientific enterprise and with the interests of the scientists and their institutions.[30]

Explaining Some of the Critiques of "Principlism" and Predicting the Result of the Debate

Another reason to describe the principles as a commensurable ethical metric is that it gives us a way of understanding the critiques of principlism and allows some tentative predictions about the challenges to be made in light of structural features of the institutions within which bioethical decisions are made. There are two types of criticism of principlism: that the principles are not a strong enough form of commensuration and that they are too strong a form of commensuration.

THE PRINCIPLES ARE NOT COMMENSURABLE ENOUGH

As I noted earlier, an ideal-type commensurable scale would not only be unitary but be numeric and interval, such as money. Risk-benefit is another commensurable scale, for example, but is not quite as precise as money due to the difficulty in quantifying risk and benefit; not that people haven't tried. The *Belmont* principles fall short of these standards. First, there are three or four of them depending on whether one is looking at the *Belmont Report* or Beauchamp and Childress's book, not one. Second, and most critically, from the perspective of perfect commensuration there is no method for comparing and weighing the principles.

Readers may recognize this as the heart of the critique of principlism by K. Danner Clouser and Bernard Gert.[31] Although I will not go into the details of their critique, what is most important for the case at hand is that the principles are not a "universal moral theory." More than one, or sometimes all, of the three or four principles are applicable to any given case, and the principles provide no mechanism for adjudicating between them. Clouser and Gert believe strongly in the "unity of morality" and further believe that "everyone must agree on the procedure to be used in deciding moral questions."[32] Why do these authors think we need a "unified theory of morality?" They conclude that "the value of using a single unified moral theory to deal with the ethical issues that arise in medicine and all other fields is that it provides a single, clear, coherent, and comprehensive decision procedure for arriving at answers."[33] In other words, decisions are more precisely calculable and predictable — a similar motivation as that for double-entry bookkeeping.

There is another critique of principlism that falls into the same camp. More accurately, this is a *concern* about the use of principles, not a critique, because it is shared by defenders of principlism. The concern is that the principle of autonomy has begun to trump all other principles.[34] I agree with critics who explain the dominance of autonomy as the result of its unusually tight articulation with American culture. I would only add that it has another attraction: it is a time-honored commensurable ethical decision-making scale that would "solve" the problem with the current principles articulated by Clouser and Gert.

One must only look at economics, with its close cousin to autonomy — individual preference — for evidence that autonomy would make a powerful commensurable metric. With all values translated into individual preference, all values can be quantified, calculated, and predicted. With this metric, economics has become a powerful field in policy making. However, one problem for bioethicists adopting autonomy as their commensurable universal metric is that their field would instantaneously come under the jurisdiction of economists, who are, after all, those who have legitimacy in the study of the aggregation of autonomous preferences for everything from automobiles to medical care.

THE PRINCIPLES ARE TOO COMMENSURABLE

The simple versions of this critique are that the values of the American people are too complex to fit the three principles described in the *Belmont Report*, or four in Beauchamp and Childress's influential text (see note 1). Perhaps twenty? Or thirty? Or perhaps particularistic principles for different racial, religious, or ethnic communities?

A more established critique, casuistry, holds that the principles are too strongly commensurable, that they fail "to give independent and sufficient attention to particular judgments about cases."[35] Instead of a "top-down" decision-making process from a few principles, casuistry would have us engage in a "bottom-up" decision making that uses as its basis analogical reasoning linking similar cases. Casuistry does not avoid principles — which are in many ways embedded in the ideal-type cases. Rather, it is opposed to the top-down application of preexisting principles.[36] It is inductive, not deductive.

Although it is unclear whether Pellegrino's "virtue theory" is a competitor to principlism,[37] it takes the position that, because the application of principles requires subjective judgment on the part of the actor, the principles require that actors have a sense of personal responsibility and integrity. In sum, principles are applied through "the judgments of sensitive and judicious persons."[38] Beauchamp, agreeing with Pellegrino, sees the theory of virtues as an important and neglected component of principles:

> Virtue theory is of the highest importance in a health-care context because a morally good person with the right motives is more likely to discern what should be done, to be motivated to do it, and to do it. The person who simply follows rules and possesses no special moral character may not be morally reliable.[39]

SOME CAUTIOUS PREDICTIONS

It is dangerous for social scientists to make predictions, but I will make a few cautious ones. My earlier discussion about the rise and success of the principles suggested that they have been successful by appearing to reduce the subjective judgment of the agents who make decisions through the rule-like qualities of their commensurable metric. Therefore, unless we foresee a decline in bureaucracy, legal influence, or government involvement in medicine and scientific research, I would predict that the ethical decision-making systems that limit the subjective judgment of actors in authority will prosper over those that give actors more interpretive leeway.

The search for a universal method of either integrating a number of principles (Clouser and Gert) or elevating one over all others (autonomy) seems more likely to succeed. The accusation that the current principles allow someone in authority (the IRB, the hospital ethics committee, etc.) to "pick whatever combination [of principles they] like" is powerful given the social constraints identified earlier.[40] Indeed, Beauchamp and Childress seem to be moving away from allowing individual subjective judgment, stating that in most recent editions of their textbook they have proposed "a decision procedure to help reduce the reliance upon intuition."[41] If forced to choose, I would suspect autonomy will become the commensurable metric for ethics. More and more instances of medical care are coming under the ultimate authority of profit-making enterprises.[42] While profit-making health maintenance organizations (HMOs), hospitals, and other institutions may maintain reliance on bioethical principles to the extent that physician professionalism remains intact,[43] the bottom line of authority in these institutions is money — the commensurable metric par excellence. With the spread of commodification[44] it seems likely that the logic of the market — autonomous choices of services controlled only by desire and financial constraint — will triumph.

To the extent that casuistry and "virtue ethics" involve greater subjective decision-making capacity on the part of actors who must ultimately legitimate their actions to government or other bureaucracies, I suspect that these schemes have little chance of success. Virtue theory seems particularly challenged here because what it is trying to revive is precisely what bureaucratic authority is attempting to remove: the reliance on the character of the individual in the office instead of "neutral" rules. I do not know enough at this point to evaluate whether virtues are normatively preferable to principles, or are a critical addition to principles, but virtues seem unlikely to be adopted by any of the institutions where health care decisions are made in modern America.

The Costs and Benefits of Principles

Let us start with the benefits of principles. To paraphrase Winston Churchill, principles are the worst system of moral decision making in existence, except for all other methods. Given the rationalized social structures within which bioethical decision making occurs, which I outlined earlier, it is hard to imagine that we could have any better. Moreover, I do not challenge the benefits of principles. First, principles acknowledge real decision-making limitations. People cannot actually consider all information in a situation, and medical students cannot be expected also to get doctoral degrees in philosophy or theology, so some information must be thrown away. But at least it should be done in an orderly fashion. Second, people do need a shared, common language to talk to each other across their particularisms.

I will briefly mention a cost of using commensurable principles that has been thoroughly addressed by people more qualified than I. Many people believe that translating complicated or "thick" values into a commensurable or "thin" metric for morality results in a severe limitation on the moral life.[45] In the HGE example that I am most familiar with, many people would probably say that there is a "principle" in relation to HGE, stating that we have a responsibility to future humans to not change their humanness. This can only be partially translated into three or four principles, and what does not get translated gets lost. More concretely, if you try to take the writings of an opponent of germ-line HGE such as Paul Ramsey — or the work of a proponent such as Robert Sinsheimer in his early days — and translate their concerns into three or four principles, many of their concerns will be left behind. The dominance of principles leaves us with James Gustafson's admonition: if we operate only on the "thin" levels of principles, then "significant issues of concern to morally sensitive persons and communities" will be left unattended.[46]

A cost of using principles that has only been gestured at by other scholars is the problem of institutionalization.[47] That is, authors have written about the "rigidity" with which the principles are applied.[48] For example, Clouser and Gert introduce their critique with the following passage: "Throughout the land, arising from the throngs of converts to bioethics awareness, there can be heard a mantra '. . . beneficence . . . autonomy . . . justice.' It is this ritual incantation in the face of biomedical dilemmas that beckons our inquiry."[49] Sociologists often write about how creating institutions devoted to the promotion of discourse — in this case, textbooks, medical school classes, academic centers, commissions — tends to result in the reification of this discourse as "fact."[50] Weber saw this as one of the most dangerous consequences of using formally rational systems, that these systems take on an "objectified, institutionalized, supra-individual form; in each sphere, rationality is embodied in the social structure and confronts individuals as something external to them."[51]

Put differently, the means come to be mistaken for the ends. For example, has the gross national product — formerly the means to the end of human happiness — become an end unto itself, maximized as it somehow represented human happiness? In the case at hand, the principles were designed to be reasonable stand-ins for the common morality or the values of society, and forwarding them was to forward this end. Yet, because we have repeated the principles so many times, the principles confront us as *the* principles of Western civilization. In careful articulations, such as those offered by Beauchamp and Childress, the principles are not considered to be inclusive of all important values. Yet in a world where texts need to be legitimate to get a hearing, questioning whether the three or four principles — now elevated to the status of ends — really represent our common morality is a risky strategy.

For example, in my study of the social determinants of changes in the debate over HGE from the 1950s to the mid-1990s, I found that the most influential texts in the 1990s discussed germ-line HGE using the principles.[52] That the most influential texts do not mention other components of society's values besides the three or four principles suggests to me that they are taken to represent morality in toto. This is especially evident when people who seem to be trying to express other concerns attempt to stuff them into the institutionalized principles. The most striking example of this is, to me, the claim that germ-line HGE is impermissible because people who do not yet exist have not given their informed consent to be experimented upon. When I have presented this argument to academics who have never heard of the principles or thought about informed consent in a serious way, they all assume that this argument was invented to fit within a highly institutionalized ethical system, akin to trying to explain the observed age of rocks when constrained within a fundamentalist Protestant theology. Put differently, before the institutionalization of principles, there were other ways of expressing this concern.

How to Limit the Cost of Principlism

During the debate about the drafting of the Constitution, Thomas Jefferson argued that we should have a revolution every nineteen years and rewrite the constitution so that the hand of the dead does not weigh too heavily on the shoulders of the living. James Madison was opposed to this notion, seeing great strength in stability.[53] If we follow Jefferson's lead, we are overdue for the revolution. If we follow Madison, perhaps we shouldn't be questioning the principles at all. As usual, perhaps the wisest course is down the middle, like the course the framers took when they created the process to amend the constitution when necessary.

I do not want to amend the principles, but instead loosen their grip on our thinking a bit. For each issue area (human experimentation, HGE, cloning, organ transplantation) a separate side debate should be started to define one principle that can be added to the accepted three or four principles. Why does this seem ridiculous to us? Partly because we have been conditioned by the institutionalization of principles to think that no more than four principles sum up moral life pretty well. It is like adding Puerto Rico to the union: Where will the fifty-first star on the flag be placed? It is also partly because we like the promise of precise commensurable metrics, not just for our economic transactions but for our ethics as well. My proposal would somewhat fragment and de-universalize bioethics. Perhaps it is too high a cost to pay. Perhaps we will lose too much in the ability to have a shared discourse and lose too much stature in the eyes of people embedded in formally rational institutions. Perhaps it leans too far toward the Jeffersonian perspective. We should, however, at least reflect on why we don't want to add another principle for each debate.

Conclusion

Weber thought that double-entry bookkeeping was "both the precursor to and the consequence of modern capitalism." Sombert went further saying, "One can say that capital, as a category, did not exist before double-entry bookkeeping."[54] Weber felt ambivalent about capitalism and the formal rationality that made it possible. The spread of formal rationality was making the world cold, spreading a sense of disenchantment. I too feel ambivalent. I do not want to forego the consumer goods produced by a capitalist system, but feel that we pay a high price in our moral life for the capitalist system.

Many scholars feel similarly about the spread of formally rational commensurable metrics in bioethics. Daniel Callahan has famously complained about people who argue for "some kind of ultimate moral big bang theory." He sees this reductionism as having reduced the ability to absorb "the insights of religion, of cultural observation and social analysis . . . and of concepts of human dignity and purpose that had a wider scope than mere autonomy."[55] It is hard to imagine going back to the pastiche of methods that reportedly comprised medical decision making before the principles. Yet, many feel that the use of principles does not capture the moral life quite properly.[56] Hopefully, by obtaining a better understanding of why the principles have such a hold upon us, we can make better decisions about how we as a society want to make our ethical decisions in medicine and science.

Notes

A shorter version of this chapter was previously published in the *Hastings Center Report* 30, no. 3 (2000): 31–38. This essay was drafted while the author was a postdoctoral fellow in the Robert Wood Johnson Health Policy Scholars program at Yale University. For a more extensive examination of the topic explored here, see John H. Evans, *Playing God? Human Genetic Engineering and the Rationalization of Public Bioethics* (Chicago: University of Chicago Press, 2002).

1. Tom L. Beauchamp and James F. Childress, *Principles of Biomedical Ethics* (New York: Oxford University Press, 1979). There have been four subsequent editions of this book.

2. Edmund D. Pellegrino, "The Metamorphosis of Medical Ethics: A Thirty-Year Retrospective," *Journal of the American Medical Association* 269 (1993): 1158–62.

3. Bruce G. Carruthers and Wendy Nelson Espeland, "Accounting for Rationality: Double-Entry Bookkeeping and the Rhetoric of Economic Rationality," *American Journal of Sociology* 97, no. 1 (1991): 31–69.

4. Ibid., 40.

5. Max Weber, *The Protestant Ethic and the Spirit of Capitalism* (New York: Charles Scribner's Sons, 1958), 181.

6. Rogers Brubaker, *The Limits of Rationality: An Essay on the Social and Moral Thought of Max Weber* (Boston: George Allen and Unwin, 1984), 44.

7. K. Danner Clouser, "Bioethics and Philosophy," *Hastings Center Report* 23, no. 6 (1993): S10–11.

8. Wendy Espeland, *The Struggle for Water: Politics, Rationality and Identity in the American Southwest* (Chicago: University of Chicago Press, 1998), 24; Wendy Nelson Espeland and Mitchell L. Stevens, "Commensuration as a Social Process," *Annual Review of Sociology* 24 (1998): 313–31.

9. Viviana A. Zelizer, *The Social Meaning of Money* (New York: Basic Books, 1994).

10. Paul Dimaggio and Walter W. Powell, "Introduction," *The New Institutionalism in Organizational Analysis*, ed. by Walter W. Powell and Paul J. DiMaggio (Chicago: University of Chicago Press, 1991), 1–40; James G. March and Herbert A. Simon, *Organizations* (New York: Wiley, 1958).

11. Espeland, *The Struggle for Water*, 25.

12. Albert Jonsen, *The Birth of Bioethics* (New York: Oxford University Press, 1998), 103.

13. Tom L. Beauchamp, "Principlism and Its Alleged Competitors," *Kennedy Institute of Ethics Journal* 5, no. 3 (1995): 181.

14. Jonsen, *The Birth of Bioethics*, 332.

15. Ibid. A recent appeal to a pluralistic common morality in the most recent edition of Beauchamp and Childress's textbook retains the commensurable metric of the principles. Instead of commensurating disparate ethical theories, the myriad values of all the citizens of the United States are commensurated into the principles.

16. Jonsen, *The Birth of Bioethics*, 333. Jeffrey Stout is less enthusiastic about this project and labels Jonsen's "common coin of moral discourse" as "moral Esperanto." Jeffrey Stout, *Ethics after Babel* (Boston: Beacon Press, 1988).

17. Many scholars have made this observation, most notably Callahan in his reference to "regulatory and policy bioethics." Daniel Callahan, "Bioethics," in *Encyclopedia of Bioethics*, 2nd ed., vol. 1, ed. Warren T. Reich (New York: Macmillan, 1995), 247–56.

18. Jonsen, *The Birth of Bioethics*, 102.

19. Ibid., 104.

20. Albert R. Jonsen, "Foreword," *A Matter of Principles? Ferment in U.S. Bioethics*, ed. Edwin R. DuBose, Ronald P. Hamel, and Laurence J. O'Connell (Valley Forge, PA: Trinity Press International, 1994), ix–xvii.

21. Yaron Ezrahi, *The Descent of Icarus: Science and the Transformation of Contemporary Democracy* (Cambridge, MA: Harvard University Press, 1990).

22. Richard Hammond, quoted in Theodore M. Porter, *Trust in Numbers: The Pursuit of Objectivity in Science and Public Life* (Princeton, NJ: Princeton University Press, 1995), 195.

23. Theodore M. Porter, *Trust in Numbers: The Pursuit of Objectivity in Science and Public Life* (Princeton, NJ: Princeton University Press, 1995), 195.

24. Ezrahi, *The Descent of Icarus*, chap. 3.

25. David J. Rothman, *Strangers by the Bedside: A History of How Law and Bioethics Transformed Medical Decision Making* (New York: Basic Books, 1991).

26. As Jonsen described the problem on the President's Commission: "One of the basic problems with the religious viewpoint is that the theologians and religious ethicists use a language which is metaphorical and symbolic very frequently.... [It is a] theological language, and non theologically trained readers have a tendency to feel it is extraordinarily vague." Quoted in Evans, *Playing God*, 116.

27. Harold I. Lief and Renee C. Fox, "Training for 'Detached Concern' in Medical Students," in *The Psychological Basis of Medical Practice*, ed. Harold I. Lief, Victor F. Lief, and Nina R. Lief (New York: Harper and Row, 1963), 12–35; Peter Conrad, "Learning to Doctor: Reflections on Recent Accounts of the Medical School Years," *Journal of Health and Social Behavior* 29 (1988): 323–32; Allen C. Smith and Sherryl Kleinman, "Managing Emotions in Medical School: Students' Contacts with the Living and the Dead," *Social Psychology Quarterly* 52, no. 1 (1989): 56–69.

28. In addition, two of the principles, beneficence and nonmaleficence, associated with the approved mechanism for their maximization, risk-benefit analysis, are a part of the ideology of scientific and medical practice.

29. Albert R. Jonsen and Lewis H. Butler, "Public Ethics and Policy Making," *Hastings Center Report* 5, no. 4 (1975): 19–31.

30. It is not that ethicists changed their beliefs to be accepted, but rather their use of principles explains why some ethicists were influential and others were not. People often find the suggestion that social forces had a hand in their success, instead of it simply being a result of their own hard work, insulting. Therefore, I will explain my own, albeit limited, success as an example of a process partially outside of my control. I was trained as a sociologist of religion, along with perhaps dozens of others in the year I entered the job market. Yet I got a job at a "prestigious research institution" while my equally smart, hard-working colleagues did not. Why? One reason is that the standards of evaluating "quality work" in the discipline select for those who have quantitative skills (which I have) and against ethnographers and interpretivists, which many of my colleagues are. I do not accept that quantitative research is necessarily "better" than ethnographic research, only that it is better given the standards that now govern my field. In the name of interdisciplinary communication, I conclude that these writers who used principles did so because they thought it was correct, not because they wanted to become influential or famous.

31. K. Danner Clouser and Bernard Gert, "A Critique of Principlism," *Journal of Medicine and Philosophy* 15 (1990): 219–36.

32. Clouser and Gert quote Mill to this effect: "There ought either to be some one fundamental principle or law at the root of all morality, or if there are several, there ought to be a determinant order of precedence among them; and

the one principle, or the rule for deciding between the various principles when they conflict, ought to be self evident." See ibid., 236.

33. Ibid., 233.

34. Paul Root Wolpe, "The Triumph of Autonomy in American Bioethics: A Sociological View," in *Bioethics and Society: Constructing the Ethical Enterprise*, ed. Raymond DeVries and Janardan Subed (Upper Saddle River, NJ: Prentice Hall, 1998), 38–59; Jonsen, *The Birth of Bioethics*, 335.

35. James F. Childress, "Ethical Theories, Principles, and Casuistry in Bioethics: An Interpretation and Defense of Principlism," in *Religious Methods and Resources in Bioethics*, ed. Paul F. Camenisch (Boston: Kluwer Academic Publishers, 1994), 190.

36. Pellegrino, "The Metamorphosis of Medical Ethics," 1161.

37. Edmund D. Pellegrino, "Toward a Virtue-Based Normative Ethics for the Health Professions," *Kennedy Institute of Ethics Journal* 5, no. 3 (1995): 253–77.

38. Beauchamp, "Principlism and Its Alleged Competitors," 193.

39. Ibid., 194–95.

40. Clouser and Gert, "A Critique of Principlism," 222.

41. Childress, "Ethical Theories, Principles, and Casuistry in Bioethics," 187.

42. Bradford H. Gray, *The Profit Motive and Patient Care: The Changing Accountability of Doctors and Hospitals* (Cambridge, MA: Harvard University Press, 1991).

43. Theodore R. Marmor, Mark Schlesinger, and Richard W. Smithey, "Nonprofit Organizations and Health Care," in *The Nonprofit Sector: A Research Handbook*, ed. Walter W. Powell (New Haven, CT: Yale University Press, 1987).

44. Margaret Radin, *Contested Commodities* (Cambridge, MA: Harvard University Press, 1996); Cass R. Sunstein, *Free Markets and Social Justice* (New York: Oxford University Press, 1997).

45. Gilbert C. Meilaender, *Body, Soul, and Bioethics* (Notre Dame, IN: University of Notre Dame Press, 1995); James M. Gustafson, "Moral Discourse about Medicine: A Variety of Forms," *The Journal of Medicine and Philosophy* 15 (1990): 125–42; Daniel Callahan, "Why America Accepted Bioethics," *The Hastings Center Report* 23 (November/December 1993): S8–S9.

46. Gustafson, "Moral Discourse about Medicine," 127.

47. But see Larry R. Churchill and Jose Jorge Siman, "Principles and the Search for Moral Certainty," *Social Science and Medicine* 23, no. 5 (1986): 463.

48. Beauchamp, "Principlism and Its Alleged Competitors," 196.

49. Clouser and Gert, "A Critique of Principlism," 219.

50. The Kennedy Institute at Georgetown educates two hundred health care professionals each year in the principles (Pellegrino, "The Metamorphosis of Medical Ethics," 1160), and Beauchamp and Childress's textbook is in its fifth edition.

51. Brubaker, *The Limits of Rationality*, 9.

52. Evans, *Playing God?*

53. Thomas Jefferson to James Madison, September 6, 1789 (cited in Joyce Appleby and Terence Ball, eds., *Thomas Jefferson: Political Writings* [Cambridge, UK: Cambridge University Press, 1999]).

54. Carruthers and Espeland, "Accounting for Rationality," 32–33.

55. Daniel Callahan, "At the Center," *Hastings Center Report* 12 (June 1982): 4.

56. Edwin R. DuBose, Ronald P. Hamel, and Laurence J. O'Connell, *A Matter of Principles? Ferment in U.S. Bioethics* (Valley Forge, PA: Trinity Press International, 1994).

Epilogue

Looking Back to Look Forward

James F. Childress

This volume looks back to look forward. It commemorates the influential *Belmont Report* by examining the following: the origins, interpretation, and application of its three principles; how they were sometimes incorporated into and employed by — and sometimes neglected by — subsequent national bodies in public bioethics; the interpretation, reinterpretation, critique, and supplementation of these principles in biomedical research and practice; and an overall assessment of methodological issues in the use of these principles, along with an examination of implications and limitations of principlism. We can appreciate the historical and enduring significance of the *Belmont Report* and its three principles without viewing them as complete and adequate formulations for the present and the future. Despite their limitations, the *Belmont* principles remain a vital part of the continuing conversation about ethical guidance for biomedical research and practice and related public policies. Yet they do not exhaust that conversation. Only a *Belmont* fundamentalist — and I have never met one — would think otherwise.

Ethical Principles in Public Bioethics: Explicit or Implicit

Sometimes the *Belmont* principles — or similar principles — appear to underlie the discussion even when they are not explicitly recognized. In response to one challenge that the National Bioethics Advisory Commission (NBAC) had not employed principles in its deliberations about human cloning, I observed

> that NBAC's concern for safety reflects the principle of nonmaleficence and that NBAC [at the time of the report] . . . could not identify benefits of human cloning that outweigh the risks to children (a consideration of beneficence) or claims of autonomy in reproduction or in scientific inquiry strong enough to outweigh the risks to children. In addition, concerns about respect for persons, including their dignity as

well as their autonomy, surfaced in discussions about objectifying and commodifying children. I would argue that these principles, and others, were transparent in NBAC's deliberation.

"At the very least," I continued, "the commission's consensus [in support of a temporary ban on human reproductive cloning] reflects its views about the respective weights of three prominent moral principles — nonmaleficence, beneficence, and respect for autonomy — in the context of recommending public policies regarding human cloning."[1]

Furthermore, principles virtually identical to the *Belmont* principles often surface as commissions reflect systematically about appropriate public policies. While the President's Commission did not explicitly employ the *Belmont* principles in its analyses and recommendations — for reasons identified in Capron's chapter — it nevertheless sometimes presented positions that appear to be based on and justified by principles very similar to the *Belmont* triad. This point holds for several of the reports of the President's Commission, as is evident in the following summary in *Deciding to Forego Life-Sustaining Treatment: Ethical, Medical, and Legal Issues in Treatment Decisions*:

> In its work on the ethical issues in health care the Commission discussed the importance of three basic values: self-determination, well-being, and equity. The concepts are not all encompassing; nor was any attempt made to relate them in a hierarchical fashion. In *Making Health Care Decisions*, the Commission focused almost entirely upon the values of self-determination and well-being, in *Securing Access to Health Care*, principally upon considerations of equity. In this Report, the Commission examines treatment situations in which all three values are intimately involved.[2]

In short, the President's Commission did not eschew general principles or values in its policy analyses, assessments, and recommendations. Despite Capron's protestations to the contrary, these "underlying values" or "basic values" — language favored by the commission — appear to be quite similar, though not identical, to the *Belmont* principles.

A public-bioethics body can achieve greater transparency in its argumentation by explicating its underlying principles. According to Leon Kass, chair of the President's Council on Bioethics, the *Belmont* principles are insufficient for bioethical guidance:

> The major principles of professional bioethics, according to the profession's own self-declaration, are these: (1) beneficence (or at least "nonmaleficence" — in plain English, "do no harm"), (2) respect for persons, and (3) justice. As applied to particular cases, these principles translate mainly into concerns to avoid bodily harm and do bodily good, to respect patient autonomy and secure informed consent, and to promote equal access to health care and provide equal protection against biohazards. So long as nobody is hurt, no one's will is violated, and no one

is excluded or discriminated against, there is little to worry about. The possibility of willing dehumanization is out of sight and out of mind.[3]

In the President's Council's first report, we find the following statement about human reproductive cloning: "If carefully considered, the concerns about safety [of cloning-to-produce-children] also begin to reveal the ethical principles that should guide a broader assessment of cloning-to-produce-children: the principles of freedom, equality, and human dignity."[4] These principles — including the concerns about safety — are at least superficially similar to the *Belmont* principles, but it is difficult to determine their content in the President's Council's report.

For instance, the principle of human dignity is central — the report is even titled *Human Cloning and Human Dignity* — and the concept of dehumanization is parasitic on that principle. However, while referred to approximately fifteen times in the report, human dignity is nowhere clearly spelled out. Hence, readers cannot easily determine whether, how, and why human reproductive cloning constitutes dehumanization. Perhaps somewhat in line with John Evans's conclusion in his essay in this volume, the President's Council seeks "thicker" bioethics than the "thin" *Belmont* principles — and related formulations — appear to provide. However, the "thick" formulations themselves require explication in order to ensure transparency in the arguments for particular policies.

Interpreting and Augmenting the *Belmont* Principles

This discussion has already identified several different formulations of general moral considerations, whether labeled principles, values, rules, or something else. As discussed by Faden, Mastroianni, and Kahn, the Advisory Committee on Human Radiation Experiments (ACHRE) formulated and drew on six basic ethical principles, which may be summarized as not treating people as mere means to others' ends, not deceiving others, not inflicting harm or risk of harm, promoting welfare and preventing harm, treating people fairly and with equal respect, and respecting others' self-determination. These principles go beyond, but also clearly overlap, the three *Belmont* principles. Whatever the formulation and number of principles, deliberation and justification in public bioethics almost invariably reflect certain core moral concerns. While the President's Commission largely ignored the *Belmont* principles — but offered some similar ones of its own — and ACHRE expanded the list to six, NBAC often used the *Belmont* principles, but still expanded and supplemented them with other moral considerations as needed.

Philosophers and others have identified several problems in the *Belmont* principles, particularly regarding beneficence and respect for persons. First, essayists in this volume differ, as do other interpreters, about whether to expand or constrict the principle of respect for persons. The problem arises because respect for persons in the *Belmont Report* encompasses two distinct kinds of actions: treating (respecting) individuals as autonomous agents and protecting persons with diminished autonomy. One response — the direction taken by Beauchamp and Childress and by Churchill, among others — is to develop a principle of respect for autonomy and to put protection of persons with diminished autonomy under beneficence. However, in Lebacqz's view, such an approach loses the original depth and breadth of respect for persons, which need to be recaptured, in part, to address oppression.

Churchill proposes a more robust principle of respect for autonomy that requires not only protecting autonomy but also promoting and enabling independent, autonomous choices. Rather than abandoning the language of respect for *persons* altogether in favor of respect for *autonomy*, Churchill proposes to place the "principle of respect for persons" in the overall, guiding vision for research involving human participants, a vision that could be stated in a preamble. It would then function as "a basic or foundational commitment," somewhat parallel to the function of the concept of human dignity in many international documents on human rights.

This discussion of the principle of respect for persons is a good example of the kinds of debates that occur about how to specify the content, range, and scope of moral principles. These debates occur with regard to other *Belmont* principles too. For example, many ethicists separate the two components of the principle of beneficence — not harming and maximizing possible benefits and minimizing possible harms — into two principles, sometimes called nonmaleficence and beneficence. One question that arises here and elsewhere is which formulation best illuminates the relevant moral concerns to guide human action.

After the *Belmont Report*, much of the discussion of respect for persons and beneficence has focused on *narrowing* their range; by contrast, much of the discussion of the principle of justice has focused on *expanding* its meaning, range, and scope. This point is evident in Patricia King's attention to procedural justice, including the participation of women and minorities in establishing agendas for research, and to compensatory justice, including coverage for injuries incurred in research. It is also evident in Susan Sherwin's use of a "distinctly feminist lens" of social justice to highlight the oppression of individuals and groups. Her analysis focuses on contexts in which individual vulnerability results largely (though not necessarily completely) from membership in oppressed social groups. These two authors, who share some other themes as well, also stress the expansion of conceptions of justice, in a changed research context, to include access to clinical trials as well as the fruits of research, in addition to protection from exploitation in research.

Individualistic versus Communitarian Principles

The debate about whether principles ought to guide biomedical research and practice has sometimes been caught in the web of the larger debate about individualism versus communitarianism. Communitarians charge that individualistic interpretations of the *Belmont* principles have dominated moral discourse and now need to be corrected. Several chapters in this volume note this deficiency — for example, Emanuel and Weijer call for attention to communities in research, Lebacqz and Sherwin draw on feminist perspectives on relationality to critique *Belmont*'s individualism, and King and Sherwin stress justice in relation to not only individuals but also groups. Changes in research (such as the increased role of genetics research) and in health care (particularly the renewed interest in public health) have undoubtedly contributed to the judgment that the *Belmont* principles are insufficient.

As noted in the Introduction, at the first NBAC meeting in October 1996, one member (Ezekiel Emanuel) contended that the three *Belmont* principles and related guidelines do not adequately address *community*. Such a challenge could mean, among other possibilities, that we should add community as a fourth principle — the approach that Emanuel and Weijer recommend in this volume — or that we should interpret all these principles through the lens of relationships and community. They argue for a principle of respect for community that generates "an obligation to respect the values and interests of the community in research and, wherever possible, to protect the community from harm." Such an independent principle, they emphasize, is needed in order to adequately recognize that communities have moral status, values, and interests that merit protection — beyond the sum of individual values and interests — and to describe and address the conflict between individual and communal interests. However, their chapter in this volume does not fully indicate how such conflicts are to be resolved, particularly in view of the different kinds of communities involved.

An alternative way to correct *Belmont*'s putative individualism could come from a richer interpretation, or reinterpretation, of its principles through the lens of relationships and community. For instance, beneficence already includes attention to contribution of knowledge to the society's welfare — this is one of the benefits to be balanced against the risks to subjects. However, attention to community might also require, as has become more common especially in genetics research, attention to harms to particular communities, such as Native American or Amish communities, beyond harms to individual members of those communities. A key component of the Emanuel-Weijer principle of respect for community is to protect communities from harm.

Reinterpreted through the lens of relationships and community, the principle of respect for persons would consider persons not merely as isolated individuals, who consent or refuse to consent to participate in research, but also as members of communities. Individuals are embedded, to varying degrees, in their communities, which have varying beliefs, values, practices, and so on. Respect for cultural values is important. However, caution is also necessary because it is not possible or justifiable to determine an individual's wishes and choices by reading them off community traditions, beliefs, and values, and it is not ethically acceptable to subordinate the individual's autonomy to the community's will without satisfying stringent justificatory conditions.

Finally, justice concerns more than fairly selecting research subjects and fairly distributing the benefits and burdens of participation in research. It may include the participation of various communities in the design and evaluation of research. Caution is needed here too, particularly to avoid the temptation to substitute community consent for individual (or surrogate) consent to participation in research. For instance, NBAC recognized that in certain contexts in international research it may be appropriate to secure permission from community representatives or from family members before approaching potential research participants. However, NBAC added the following restrictions: "in no case may permission from a community representative or council replace the requirement of a competent individual's voluntary informed consent," and researchers "should strive to ensure that individuals agree to participate in research without coercion or undue inducements from community leaders or representatives." Similarly, "in no case . . . may a family member's permission replace the requirement of a competent individual's voluntary informed consent."[5]

Beyond community participation in the design and evaluation of research, justice should include, as King argues, compensation for research-related injuries. Compensation can also be viewed as an expression of the community's solidarity with research participants who assume a position of risk on behalf of the community and are nonnegligently injured in the process. From this standpoint, it is not sufficient to disclose on the consent form — as the Common Rule requires — whether any compensation will be provided for research-related injuries that are nonnegligently caused. Instead, compensation should be provided as a matter of moral obligation. Countries with a commitment to universal access to health care generally do not view some compensation for research-related injuries as a problem — such injuries would be routinely covered, at least for medical expenses. However, there are always questions about the scope of compensation, including whether a compensation mechanism should cover, for example, lost wages, disability, and death, in addition to immediate medical problems.

As these examples suggest, it is both possible and important to revisit the *Belmont* principles in light of concerns about relationships and community. This does not require, in my judgment, a new principle of respect for community, as Emanuel and Weijer propose. Instead, a broader and richer interpretation of the principles is sufficient. Reinterpreting these principles through the lens of relationships and community can be illuminating, as long as these reinterpretations do not ignore or distort what has been important in more individualistic interpretations. In addition, as feminist interpreters remind us, it is important to attend to the oppression of individuals in various relationships and communities along with the oppression of groups in which they participate.

Other Moral Concepts and Foundations

How are the *Belmont* principles distinguished from and related to other formulations of principles for guiding biomedical research and practice? While some essays in this volume distinguish and relate the *Belmont* principles to Beauchamp and Childress's four principles of respect for autonomy, beneficence, nonmaleficence, and justice, a number of other formulations could also be explored. More specifically, it would be useful to examine the similarities and differences between the *Belmont* principles, often taken to characterize the U.S. ethical framework for addressing research involving human subjects, and evolving European frameworks. One formulation of "Basic Ethical Principles in European Bioethics and Biolaw" identifies autonomy, dignity, integrity, and vulnerability within a social justice framework of solidarity and responsibility. Another framework that appears in many international formulations of principles of biomedical research and practice is human rights, for which, in many documents, human dignity is the foundation. In the context of globalization, attention to the human-rights framework is imperative.

Belmont presupposes a particular context: "Three basic principles, among those generally accepted in our cultural tradition, are particularly relevant to the ethics of research involving human subjects." Regarding the scope of these principles, several essays stress, as we have seen, the applicability of these principles — or similar principles — outside the range of research. One example is Cassell's discussion of their embodiment in the ethos and practice of medicine. In addition, there is the question of their connection with, and limitation to, "our cultural tradition." This question becomes particularly important in the context of international research that the United States sponsors or U.S. researchers conduct in developing countries — a context addressed by NBAC in one of its reports. One fundamental question is whether *Belmont* represents cultural imperialism or, instead, a set of moral concerns that are broadly recognized and applicable, even if expressed in different language, such as human rights, and specified differently.

Attention to ethical principles has sometimes obscured other relevant moral concepts in practice. For instance, Faden, Mastroianni, and Kahn observe that the concept of trust played an important role in ACHRE's analyses and deliberations, even though it is not captured in most lists of principles. They further contend that trust depends on transparency of argumentation as well as on a perception of fundamental commitments, which may be articulated in the form of principles.

Principlism: Possibilities and Limitations

The term "principlism" is a pejorative label for several positions, including the *Belmont* principles. According to critics, principlism employs several intermediate moral principles without grounding them in a theory and without providing a rank ordering. The *Belmont* principles thus appear within a principlist framework that does not itself indicate how any conflicts among the three principles can and should be resolved.

In this volume, Veatch and Richardson both offer interpretations and critiques of the *Belmont* principlist framework, as well as other versions of principlism. And each offers a distinctive approach to the avoidance, reduction, or resolution of conflict among these principles. Richardson further develops his influential model of specification, a process that is more tightly defined than the specification, concretization, or interpretation that already occurs in the *Belmont Report* itself. And Veatch further explicates a complex model that includes a rank-order, coupled with limited balancing.

These approaches address, in different ways, two dimensions of ethical principles and rules: on the one hand, their meaning, range, and scope, and, on the other hand, their weight, strength, or stringency. Richardson's model of specification mainly focuses on the former, whereas Veatch's rank-order combined with balancing mainly focuses on the latter. We can expect vigorous debates to continue about how to specify ethical principles and about how to resolve their conflicts if specification does not succeed in avoiding or reducing conflicts.

In short, looking back at the *Belmont* principles is a way to look — and to move — forward. We look back not because we suppose that the *Belmont* principles are adequate for contemporary biomedical research and practice but instead to recognize their historical and contemporary significance by engaging and interrogating them in current conversations about bioethical guidance. We can learn much from various efforts to apply, modify, expand, restrict, supplement, and reinterpret these principles as we seek to determine the best framework for analyzing and directing biomedical research and practice.

Notes

1. James F. Childress, "The Challenge of Public Ethics: Reflections on NBAC's Report," *Hastings Center Report* 27, no. 5 (1997): 10.

2. President's Commission for the Study of Ethical Problems in Medicine and Biomedical and Behavioral Research, *Deciding to Forego Life-Sustaining Treatment: A Report on the Ethical, Medical, and Legal Issues in Treatment Decisions* (Washington, DC: U.S. Government Printing Office, March 1983), 26.

3. Leon R. Kass, *Life, Liberty and the Defense of Dignity: The Challenge for Bioethics* (San Francisco, CA: Encounter Books, 2002), vi.

4. President's Council on Bioethics, *Human Cloning and Human Dignity* (New York: Public Affairs, 2002), xlvii.

5. National Bioethics Advisory Commission, *Ethical and Policy Issues in International Research: Clinical Trials in Developing Countries*, vol. 1, *Report and Recommendations of the National Bioethics Advisory Commission* (Bethesda, MD: NBAC, 2001), vi–vii, 42–45.

APPENDIX

The Belmont Report

Office of the Secretary

Ethical Principles and Guidelines for the Protection
of Human Subjects of Research

The National Commission for the Protection of Human Subjects
of Biomedical and Behavioral Research

April 18, 1979

AGENCY: Department of Health, Education, and Welfare.

ACTION: Notice of Report for Public Comment.

SUMMARY: On July 12, 1974, the National Research Act (Pub. L. 93-348) was signed into law, there-by creating the National Commission for the Protection of Human Subjects of Biomedical and Behavioral Research. One of the charges to the Commission was to identify the basic ethical principles that should underlie the conduct of biomedical and behavioral research involving human subjects and to develop guidelines which should be followed to assure that such research is conducted in accordance with those principles. In carrying out the above, the Commission was directed to consider: (i) the boundaries between biomedical and behavioral research and the accepted and routine practice of medicine, (ii) the role of assessment of risk-benefit criteria in the determination of the appropriateness of research involving human subjects, (iii) appropriate guidelines for the selection of human subjects for participation in such research and (iv) the nature and definition of informed consent in various research settings.

The Belmont Report attempts to summarize the basic ethical principles identified by the Commission in the course of its deliberations. It is the outgrowth of an intensive four-day period of discussions that were held in February 1976 at the Smithsonian Institution's Belmont Conference Center supplemented by the monthly deliberations of the Commission that were held over a period of nearly four years. It is a statement of basic ethical principles and guidelines that should assist in resolving the ethical problems that surround the conduct of research with human subjects. By publishing the Report in the *Federal Register,* and providing reprints upon request, the Secretary intends that it may be made readily available to scientists, members of Institutional Review Boards, and Federal employees. The two-volume Appendix, containing the lengthy reports of experts and specialists who assisted the Commission in fulfilling this part of its charge, is available as DHEW Publication No. (OS) 78-0013 and No. (OS) 78-0014, for sale by the Superintendent of Documents, U.S. Government Printing Office, Washington, D.C. 20402.

Unlike most other reports of the Commission, the Belmont Report does not make specific recommendations for administrative action by the Secretary of Health, Education, and Welfare. Rather, the Commission recommended that the Belmont Report be adopted in its entirety, as a statement of the Department's policy. The Department requests public comment on this recommendation.

National Commission for the Protection of Human Subjects of Biomedical and Behavioral Research

MEMBERS OF THE COMMISSION

Kenneth John Ryan, M.D., Chairman, Chief of Staff,
Boston Hospital for Women.

Joseph V. Brady, Ph.D., Professor of Behavioral Biology,
Johns Hopkins University.

Robert E. Cooke, M.D., President,
Medical College of Pennsylvania.

Dorothy I. Height, President,
National Council of Negro Women, Inc.

Albert R. Jonsen, Ph.D., Associate Professor of Bioethics,
University of California at San Francisco.

Patricia King, J.D., Associate Professor of Law,
Georgetown University Law Center.

Karen Lebacqz, Ph.D., Associate Professor of Christian Ethics,
Pacific School of Religion.

***David W. Louisell, J.D., Professor of Law,
University of California at Berkeley.

Donald W. Seldin, M.D., Professor and Chairman,
Department of Internal Medicine, University of Texas at Dallas.

***Eliot Stellar, Ph.D.,
Provost of the University and Professor of Physiological Psychology,
University of Pennsylvania.

***Robert H. Turtle, LL.B., Attorney,
VomBaur, Coburn, Simmons & Turtle, Washington, D.C.

****Deceased.*

TABLE OF CONTENTS

Ethical Principles & Guidelines for Research Involving Human Subjects

Scientific research has produced substantial social benefits. It has also posed some troubling ethical questions. Public attention was drawn to these questions by reported abuses of human subjects in biomedical experiments, especially during the Second World War. During the Nuremberg War Crime Trials, the Nuremberg code was drafted as a set of standards for judging physicians and scientists who had conducted biomedical experiments on concentration camp prisoners. This code became the prototype of many later codes[1] intended to assure that research involving human subjects would be carried out in an ethical manner.

The codes consist of rules, some general, others specific, that guide the investigators or the reviewers of research in their work. Such rules often are inadequate to cover complex situations; at times they come into conflict, and they are frequently difficult to interpret or apply. Broader ethical principles will provide a basis on which specific rules may be formulated, criticized and interpreted.

Three principles, or general prescriptive judgments, that are relevant to research involving human subjects are identified in this statement. Other principles may also be relevant. These three are comprehensive, however, and are stated at a level of generalization that should assist scientists, subjects, reviewers and interested citizens to understand the ethical issues inherent in research involving human subjects. These principles cannot always be applied so as to resolve beyond dispute particular ethical problems. The objective is to provide an analytical framework that will guide the resolution of ethical problems arising from research involving human subjects.

This statement consists of a distinction between research and practice, a discussion of the three basic ethical principles, and remarks about the application of these principles.

Part A: Boundaries between Practice & Research

A. BOUNDARIES BETWEEN PRACTICE AND RESEARCH

It is important to distinguish between biomedical and behavioral research, on the one hand, and the practice of accepted therapy on the other, in order to know what activities ought to undergo review for the protection of human subjects of research. The distinction between research and practice is blurred partly because both often occur together (as in research designed to evaluate a therapy) and partly because notable departures from standard practice are often called "experimental" when the terms "experimental" and "research" are not carefully defined.

For the most part, the term "practice" refers to interventions that are designed solely to enhance the well-being of an individual patient or client and that have a reasonable expectation of success. The purpose of medical or behavioral practice is to provide diagnosis, preventive treatment or therapy to particular individuals.[2] By contrast, the term "research" designates an activity designed to test an hypothesis, permit conclusions to be drawn, and thereby to develop or contribute to generalizable knowledge (expressed, for example, in theories, principles, and statements of relationships). Research is usually described in a formal protocol that sets forth an objective and a set of procedures designed to reach that objective.

When a clinician departs in a significant way from standard or accepted practice, the innovation does not, in and of itself, constitute research. The fact that a procedure is "experimental," in the sense of new, untested or different, does not automatically place it in the category of research. Radically new procedures of this description should, however, be made the object of formal research at an early stage in order to determine whether they are safe and effective. Thus, it is the responsibility of medical practice committees, for example, to insist that a major innovation be incorporated into a formal research project.[3]

Research and practice may be carried on together when research is designed to evaluate the safety and efficacy of a therapy. This need not cause any confusion regarding whether or not the activity requires review; the general rule is that if there is any element of research in an activity, that activity should undergo review for the protection of human subjects.

Part B: Basic Ethical Principles

B. BASIC ETHICAL PRINCIPLES

The expression "basic ethical principles" refers to those general judgments that serve as a basic justification for the many particular ethical prescriptions and evaluations of human actions. Three basic principles, among those generally accepted in our cultural tradition, are particularly relevant to the ethics of research involving human subjects: the principles of respect of persons, beneficence and justice.

1. Respect for Persons.

Respect for persons incorporates at least two ethical convictions: first, that individuals should be treated as autonomous agents, and second, that persons with diminished autonomy are entitled to protection. The principle of respect for persons thus divides into two separate moral requirements: the requirement to acknowledge autonomy and the requirement to protect those with diminished autonomy.

An autonomous person is an individual capable of deliberation about personal goals and of acting under the direction of such deliberation. To respect autonomy is to give weight to autonomous persons' considered opinions and choices while refraining from obstructing their actions unless they are clearly detrimental to others. To show lack of respect for an autonomous agent is to repudiate that person's considered judgments, to deny an individual the freedom to act on those considered judgments, or to withhold information necessary to make a considered judgment, when there are no compelling reasons to do so.

However, not every human being is capable of self-determination. The capacity for self-determination matures during an individual's life, and some individuals lose this capacity wholly or in part because of illness, mental disability, or circumstances that severely restrict liberty. Respect for the immature and the incapacitated may require protecting them as they mature or while they are incapacitated.

Some persons are in need of extensive protection, even to the point of excluding them from activities which may harm them; other persons require little protection beyond making sure they undertake activities freely and with awareness of possible adverse consequence. The extent of protection afforded should depend upon the risk of harm and the likelihood of benefit. The judgment that any individual lacks autonomy should be periodically reevaluated and will vary in different situations.

In most cases of research involving human subjects, respect for persons demands that subjects enter into the research voluntarily and with adequate information. In some situations, however, application of the principle is not obvious. The involvement of prisoners as subjects of research provides an instructive example. On the one hand, it would seem that the principle of respect for persons requires that prisoners not be deprived of the opportunity to volunteer for research. On the other hand, under prison conditions they may be subtly coerced or unduly influenced to engage in research activities for which they would not otherwise volunteer. Respect for persons would then dictate that prisoners be protected. Whether to allow prisoners to "volunteer" or to "protect" them presents a dilemma. Respecting persons, in most hard cases, is often a matter of balancing competing claims urged by the principle of respect itself.

2. Beneficence.

Persons are treated in an ethical manner not only by respecting their decisions and protecting them from harm, but also by making efforts to secure their well-being. Such treatment falls under the principle of beneficence. The term "beneficence" is often understood to cover acts of kindness or charity that go beyond strict obligation. In this document, beneficence is understood in a stronger sense, as an obligation. Two general rules have been formulated as complementary expressions of beneficent actions in this sense: (1) do not harm and (2) maximize possible benefits and minimize possible harms.

The Hippocratic maxim "do no harm" has long been a fundamental principle of medical ethics. Claude Bernard extended it to the realm of research, saying that one should not injure one person regardless of the benefits that might come to others. However, even avoiding harm requires learning what is harmful; and, in the process of obtaining this information, persons may be exposed to risk of harm. Further, the Hippocratic Oath requires physicians to benefit their patients "according to their best judgment." Learning what will in fact benefit may require exposing persons to risk. The problem posed by these imperatives is to decide when it is justifiable to seek certain benefits despite the risks involved, and when the benefits should be foregone because of the risks.

The obligations of beneficence affect both individual investigators and society at large, because they extend both to particular research projects and to the entire enterprise of research. In the case of particular projects, investigators and members of their institutions are obliged to give forethought to the maximization of benefits and the reduction of risk that might occur from the research investigation. In the case of scientific research in general, members of the larger society are obliged to recognize the longer term benefits and risks that may result from the improvement of knowledge and from the development of novel medical, psychotherapeutic, and social procedures.

The principle of beneficence often occupies a well-defined justifying role in many areas of research involving human subjects. An example is found in research involving children. Effective ways of treating childhood diseases and fostering healthy development are benefits that serve to justify research involving children — even when individual research subjects are not direct beneficiaries. Research also makes it possible to avoid the harm that may result from the application of previously accepted routine practices that on closer investigation turn out to be dangerous. But the role of the principle of beneficence is not always so unambiguous. A difficult ethical problem remains, for example, about research that presents more than minimal risk without immediate prospect of direct benefit to the children involved. Some have argued that such research is inadmissible, while others have pointed out that this limit would rule out much research promising great benefit to children in the future. Here again, as with all hard cases, the different claims covered by the principle of beneficence may come into conflict and force difficult choices.

3. Justice.

Who ought to receive the benefits of research and bear its burdens? This is a question of justice, in the sense of "fairness in distribution" or "what is deserved." An injustice occurs when some benefit to which a person is entitled is denied without good reason or when some burden is imposed unduly. Another way of conceiving the principle of justice is that equals ought to be treated equally. However, this statement requires explication. Who is equal and who is unequal? What considerations justify departure from equal distribution? Almost all commentators allow that distinctions based on experience, age, deprivation, competence, merit and position do sometimes constitute criteria justifying differential treatment for certain purposes. It is necessary, then, to explain in what respects people should be treated equally. There are several widely accepted formulations of just ways to distribute burdens and benefits. Each formulation mentions some relevant property on the basis of which burdens and benefits should be distributed. These formulations are (1) to each person an equal share, (2) to each person according to individual need, (3) to each person according to individual effort, (4) to each person according to societal contribution, and (5) to each person according to merit.

Questions of justice have long been associated with social practices such as punishment, taxation and political representation. Until recently these questions have not generally been associated with scientific research. However, they are foreshadowed even in the earliest reflections on the ethics of research involving human subjects. For example, during the 19th and early 20th centuries the burdens of serving as research subjects fell largely upon poor ward patients, while the benefits of improved medical care flowed primarily to private patients. Subsequently, the exploitation of unwilling prisoners as research subjects in Nazi concentration camps was condemned as a particularly flagrant injustice. In this country, in the 1940s, the Tuskegee syphilis study used disadvantaged, rural black men to study the untreated course of a disease that is by no means confined to that population. These subjects were deprived of demonstrably effective treatment in order not to interrupt the project, long after such treatment became generally available.

Against this historical background, it can be seen how conceptions of justice are relevant to research involving human subjects. For example, the selection of research subjects needs to be scrutinized in order to determine whether some classes (e.g., welfare patients, particular racial and ethnic minorities, or persons confined to institutions) are being systematically selected simply because of their easy availability, their compromised position, or their manipulability, rather than for reasons directly related to the problem being studied. Finally, whenever research supported by public funds leads to the development of therapeutic devices and procedures, justice demands both that these not provide advantages only to those who can afford them and that such research should not unduly involve persons from groups unlikely to be among the beneficiaries of subsequent applications of the research.

C. APPLICATIONS

Applications of the general principles to the conduct of research leads to consideration of the following requirements: informed consent, risk/benefit assessment, and the selection of subjects of research.

1. Informed Consent.

Respect for persons requires that subjects, to the degree that they are capable, be given the opportunity to choose what shall or shall not happen to them. This opportunity is provided when adequate standards for informed consent are satisfied.

While the importance of informed consent is unquestioned, controversy prevails over the nature and possibility of an informed consent. Nonetheless, there is widespread agreement that the consent process can be analyzed as containing three elements: information, comprehension and voluntariness.

Information. Most codes of research establish specific items for disclosure intended to assure that subjects are given sufficient information. These items generally include: the research procedure, their purposes, risks and anticipated benefits, alternative procedures (where therapy is involved), and a statement offering the subject the opportunity to ask questions and to withdraw at any time from the research. Additional items have been proposed, including how subjects are selected, the person responsible for the research, etc.

However, a simple listing of items does not answer the question of what the standard should be for judging how much and what sort of information should be provided. One standard frequently invoked in medical practice, namely the information commonly provided by practitioners in the field or in the locale, is inadequate since research takes place precisely when a common understanding does not exist. Another standard, currently popular in malpractice law, requires the practitioner to reveal the information that reasonable persons would wish to know in order to make a decision regarding their care. This, too, seems insufficient since the research subject, being in essence a volunteer, may wish to know considerably more about risks gratuitously undertaken than do patients who deliver themselves into the hand of a clinician for needed care. It may be that a standard of "the reasonable volunteer" should be proposed: the extent and nature of information should be such that persons, knowing that the procedure is neither necessary for their care nor perhaps fully understood, can decide whether they wish to participate in the furthering of knowledge. Even when some direct benefit to them is anticipated, the subjects should understand clearly the range of risk and the voluntary nature of participation.

A special problem of consent arises where informing subjects of some pertinent aspect of the research is likely to impair the validity of the research. In many cases, it is sufficient to indicate to subjects that they are being invited to participate in research of which some features will not be revealed until the research is concluded. In all cases of research involving incomplete disclosure, such research is justified only if it is clear that (1) incomplete disclosure is truly necessary to accomplish the goals of the research, (2) there are no undisclosed risks to subjects that are more than minimal, and (3) there is an adequate plan for debriefing subjects, when appropriate, and for dissemination of research results to them. Information about risks should never be withheld for the purpose of eliciting the cooperation of subjects, and truthful answers should always be given to direct questions about the research. Care should be taken to distinguish cases in which disclosure would destroy or invalidate the research from cases in which disclosure would simply inconvenience the investigator.

Comprehension. The manner and context in which information is conveyed is as important as the information itself. For example, presenting information in a disorganized and rapid fashion, allowing too little time for consideration or curtailing opportunities for questioning, all may adversely affect a subject's ability to make an informed choice.

Because the subject's ability to understand is a function of intelligence, rationality, maturity and language, it is necessary to adapt the presentation of the information to the subject's capacities. Investigators are responsible for ascertaining that the subject has comprehended the information. While there is always an obligation to ascertain that the information about risk to subjects is complete and adequately comprehended, when the risks are more serious, that obligation increases. On occasion, it may be suitable to give some oral or written tests of comprehension.

Special provision may need to be made when comprehension is severely limited — for example, by conditions of immaturity or mental disability. Each class of subjects that one might consider as incompetent (e.g., infants and young children, mentally disable patients, the terminally ill and the comatose) should be considered on its own terms. Even for these persons, however, respect requires giving them the opportunity to choose, to the extent they are able, whether or not to participate in research. The objections of these subjects to involvement should be honored, unless the research entails providing them a therapy unavailable elsewhere. Respect for persons also requires seeking the permission of other parties in order to protect the subjects from harm. Such persons are thus respected both by acknowledging their own wishes and by the use of third parties to protect them from harm.

The third parties chosen should be those who are most likely to understand the incompetent subject's situation and to act in that person's best interest. The person authorized to act on behalf of the subject should be given an opportunity to observe the research as it proceeds in order to be able to withdraw the subject from the research, if such action appears in the subject's best interest.

Voluntariness. An agreement to participate in research constitutes a valid consent only if voluntarily given. This element of informed consent requires conditions free of coercion and undue influence. Coercion occurs when an overt threat of harm is intentionally presented by one person to another in order to obtain compliance. Undue influence, by contrast, occurs through an offer of an excessive, unwarranted, inappropriate or improper reward or other overture in order to obtain compliance. Also, inducements that would ordinarily be acceptable may become undue influences if the subject is especially vulnerable.

Unjustifiable pressures usually occur when persons in positions of authority or commanding influence — especially where possible sanctions are involved — urge a course of action for a subject. A continuum of such influencing factors exists, however, and it is impossible to state precisely where justifiable persuasion ends and undue influence begins. But undue influence would include actions such as manipulating a person's choice through the controlling influence of a close relative and threatening to withdraw health services to which an individual would otherwise be entitled.

2. Assessment of Risks and Benefits.

The assessment of risks and benefits requires a careful arrayal of relevant data, including, in some cases, alternative ways of obtaining the benefits sought in the research. Thus, the assessment presents both an opportunity and a responsibility to gather systematic and comprehensive information about proposed research. For the investigator, it is a means to examine whether the proposed research is properly designed. For a review committee, it is a method for determining whether the risks that will be presented to subjects are justified. For prospective subjects, the assessment will assist the determination whether or not to participate.

The Nature and Scope of Risks and Benefits. The requirement that research be justified on the basis of a favorable risk/benefit assessment bears a close relation to the principle of beneficence, just as the moral requirement that informed consent be obtained is derived primarily from the principle of respect for persons. The term "risk" refers to a possibility that harm may occur. However, when expressions such as "small risk" or "high risk" are used, they usually refer (often ambiguously) to both the chance (probability) of experiencing a harm and the severity (magnitude) of the envisioned harm.

The term "benefit" is used in the research context to refer to something of positive value related to health or welfare. Unlike "risk," "benefit" is not a term that expresses probabilities. Risk is properly contrasted to probability of benefits, and benefits are properly contrasted with harms rather than risks of harm. Accordingly, so-called risk/benefit assessments are concerned with the probabilities and magnitudes of possible harm and anticipated benefits. Many kinds of possible harms and benefits need to be taken into account. There are, for example, risks of psychological harm, physical harm, legal harm, social harm and economic harm and the corresponding benefits. While the most likely types of harms to research subjects are those of psychological or physical pain or injury, other possible kinds should not be overlooked.

Risks and benefits of research may affect the individual subjects, the families of the individual subjects, and society at large (or special groups of subjects in society). Previous codes and Federal regulations have required that risks to subjects be outweighed by the sum of both the anticipated benefit to the subject, if any, and the anticipated benefit to society in the form of knowledge to be gained from the research. In balancing these different elements, the risks and benefits affecting the immediate research subject will normally carry special weight. On the other hand, interests other than those of the subject may on some occasions be sufficient by themselves to justify the risks involved in the research, so long as the subject's rights have been protected. Beneficence thus requires that we protect against risk of harm to subjects and also that we be concerned about the loss of the substantial benefits that might be gained from research.

The Systematic Assessment of Risks and Benefits. It is commonly said that benefits and risks must be "balanced" and shown to be "in a favorable ratio." The metaphorical character of these terms draws attention to the difficulty of making precise judgments. Only on rare occasions will quantitative techniques be available for the scrutiny of research protocols. However, the idea of systematic, nonarbitrary analysis of risks and benefits should be emulated insofar as possible. This ideal requires those making decisions about the justifiability of research to be thorough in the accumulation and assessment of information about all aspects of the research, and to consider alternatives systematically. This procedure renders the assessment of research more rigorous and precise, while making communication between review board members and investigators less subject to misinterpretation, misinformation and conflicting judgments. Thus, there should first be a determination of the validity of the presuppositions of the research; then the nature, probability and magnitude of risk should be distinguished with as much clarity as possible. The method of ascertaining risks should be explicit, especially where there is no alternative to the use of such vague categories as small or slight risk. It should also be determined whether an investigator's estimates of the probability of harm or benefits are reasonable, as judged by known facts or other available studies.

Finally, assessment of the justifiability of research should reflect at least the following considerations: **(i)** Brutal or inhumane treatment of human subjects is never morally justified. **(ii)** Risks should be reduced to those necessary to achieve the research objective. It should be determined whether it is in fact necessary to use human subjects at all. Risk can perhaps never be entirely eliminated, but it can often be reduced by careful attention to alternative procedures. **(iii)** When research involves significant risk of serious impairment, review committees should be extraordinarily insistent on the justification of the risk (looking usually to the likelihood of benefit to the subject — or, in some rare cases, to the manifest voluntariness of the participation). **(iv)** When vulnerable populations are involved in research, the appropriateness of involving them should itself be demonstrated. A number of variables go into such judgments, including the nature and degree of risk, the condition of the particular population involved, and the nature and level of the anticipated benefits. **(v)** Relevant risks and benefits must be thoroughly arrayed in documents and procedures used in the informed consent process.

3. Selection of Subjects.

Just as the principle of respect for persons finds expression in the requirements for consent, and the principle of beneficence in risk/benefit assessment, the principle of justice gives rise to moral requirements that there be fair procedures and outcomes in the selection of research subjects.

Justice is relevant to the selection of subjects of research at two levels: the social and the individual. Individual justice in the selection of subjects would require that researchers exhibit fairness: thus, they should not offer potentially beneficial research only to some patients who are in their favor or select only "undesirable" persons for risky research. Social justice requires that distinction be drawn between classes of subjects that ought, and ought not, to participate in any particular kind of research, based on the ability of members of that class to bear burdens and on the appropriateness of placing further burdens on already burdened persons. Thus, it can be considered a matter of social justice that there is an order of preference in the selection of classes of subjects (e.g., adults before children) and that some classes of potential subjects (e.g., the institutionalized mentally infirm or prisoners) may be involved as research subjects, if at all, only on certain conditions.

Injustice may appear in the selection of subjects, even if individual subjects are selected fairly by investigators and treated fairly in the course of research. Thus injustice arises from social, racial, sexual and cultural biases institutionalized in society. Thus, even if individual researchers are treating their research subjects fairly, and even if IRBs are taking care to assure that subjects are selected fairly within a particular institution, unjust social patterns may nevertheless appear in the overall distribution of the burdens and benefits of research. Although individual institutions or investigators may not be able to resolve a problem that is pervasive in their social setting, they can consider distributive justice in selecting research subjects.

Some populations, especially institutionalized ones, are already burdened in many ways by their infirmities and environments. When research is proposed that involves risks and does not include a therapeutic component, other less burdened classes of persons should be called upon first to accept these risks of research, except where the research is directly related to the specific conditions of the class involved. Also, even though public funds for research may often flow in the same directions as public funds for health care, it seems unfair that populations dependent on public health care constitute a pool of preferred research subjects if more advantaged populations are likely to be the recipients of the benefits.

One special instance of injustice results from the involvement of vulnerable subjects. Certain groups, such as racial minorities, the economically disadvantaged, the very sick, and the institutionalized, may continually be sought as research subjects, owing to their ready availability in settings where research is conducted. Given their dependent status and their frequently compromised capacity for free consent, they should be protected against the danger of being involved in research solely for administrative convenience, or because they are easy to manipulate as a result of their illness or socioeconomic condition.

Notes

1. Since 1945, various codes for the proper and responsible conduct of human experimentation in medical research have been adopted by different organizations. The best known of these codes are the Nuremberg Code of 1947, the Helsinki Declaration of 1964 (revised in 1975), and the 1971 Guidelines (codified into Federal Regulations in 1974) issued by the U.S. Department of

Health, Education, and Welfare. Codes for the conduct of social and behavioral research have also been adopted, the best known being that of the American Psychological Association, published in 1973.

2. Although practice usually involves interventions designed solely to enhance the well-being of a particular individual, interventions are sometimes applied to one individual for the enhancement of the well-being of another (e.g., blood donation, skin grafts, organ transplants), or an intervention may have the dual purpose of enhancing the well-being of a particular individual, and, at the same time, providing some benefit to others (e.g., vaccination, which protects both the person who is vaccinated and society generally). The fact that some forms of practice have elements other than immediate benefit to the individual receiving an intervention, however, should not confuse the general distinction between research and practice. Even when a procedure applied in practice may benefit some other person, it remains an intervention designed to enhance the well-being of a particular individual or groups of individuals; thus, it is practice and need not be reviewed as research.

3. Because the problems related to social experimentation may differ substantially from those of biomedical and behavioral research, the Commission specifically declines to make any policy determination regarding such research at this time. Rather, the Commission believes that the problem ought to be addressed by one of its successor bodies.

Contributors

TOM L. BEAUCHAMP is a professor of philosophy at Georgetown University and a senior research scholar at the Kennedy Institute of Ethics. He is the author of numerous books in ethics, including *Principles of Biomedical Ethics*, 5th ed., co-authored with James Childress. He served on the National Commission for the Protection of Human Subjects of Biomedical and Behavioral Research.

ALEXANDER M. CAPRON is University Professor, Henry W. Bruce Professor of Equity (The Law School), professor of law and medicine at the Keck School of Medicine, and director of the Pacific Center for Health Policy and Ethics, University of Southern California. He is currently on leave as director of the Department of Ethics, Trade, Human Rights and Health Law, World Health Organization, Geneva. He was executive director of the President's Commission and a member of the National Bioethics Advisory Commission.

ERIC J. CASSELL is clinical professor of public health at Cornell University Medical College in New York City. He is the author of several books, including *The Healer's Art* (1985), *The Nature of Suffering and the Goals of Medicine* (1994), and *Doctoring: The Nature of Primary Care Medicine* (1997). He was a member of the National Bioethics Advisory Commission.

JAMES F. CHILDRESS is the John Allen Hollingsworth Professor of Ethics and director of the Institute for Practical Ethics and Public Life at the University of Virginia. His several books include *Principles of Biomedical Ethics*, 5th ed., co-authored with Tom L. Beauchamp. Childress was a consultant to the National Commission for the Protection of Human Subjects and a member of the National Bioethics Advisory Commission.

LARRY R. CHURCHILL is the Ann Geddes Stahlman Professor of Medical Ethics at the Vanderbilt University Medical Center in Nashville. He is the author of *Rationing Health Care in America: Perceptions and Principles of Justice* (1987) and *Self-Interest and Universal Health Care: Why Well-Insured Americans Should Support Coverage for Everyone* (1994) and editor of *The Social Medicine Reader* (1997).

EZEKIEL J. EMANUEL is chair of the Department of Clinical Bioethics, Warren G. Magnuson Clinical Center, National Institutes of Health (NIH). He is the author of *The Ends of Human Life: Medical Ethics in a Liberal Polity* (1992) and coeditor of *Ethical and Regulatory Aspects of Clinical Research: Readings and Commentary* (2004). He was a member of the National Bioethics Advisory Commission until his appointment to NIH.

JOHN H. EVANS is associate professor of sociology at the University of California, San Diego. He is the author of *Playing God? Human Genetic Engineering and the Rationalization of Public Bioethical Debate* (2002) and coeditor with Robert Wuthnow of *The Quiet Hand of God: Faith-Based Activism and the Public Role of Mainline Protestantism* (2002).

RUTH R. FADEN is the Philip Franklin Wagley Professor of Biomedical Ethics and executive director of The Phoebe R. Berman Bioethics Institute at Johns Hopkins University. Among her many publications, she is coauthor with Tom L. Beauchamp of *A History and Theory of Informed Consent* (1986). Faden chaired the presidential Advisory Committee on Human Radiation Experiments.

ALBERT R. JONSEN taught at the University of California School of Medicine, San Francisco, and at the University of Washington. His several books include *The Birth of Bioethics* (1998). He served on the National Commission for the Protection of Human Subjects of Biomedical and Behavioral Research and on the President's Commission for the Study of Ethical Problems in Medicine and Biomedical and Behavioral Research.

JEFFREY P. KAHN, who holds the Maas Family Endowed Chair in Bioethics, is director, Center for Bioethics, University of Minnesota. The author of numerous articles in bioethics, he is coeditor of *Beyond Consent: Seeking Justice in Research* (1998). He served as associate director of the White House Advisory Committee on Human Radiation Experiments in 1994–95.

PATRICIA A. KING is the Carmack Waterhouse Professor of Law, Medicine, Ethics, and Public Policy at Georgetown Law Center. The coauthor of *Law, Science and Medicine* (3rd ed., 2005), King was a member of the National Commission for the Protection of Human Subjects of Biomedical and Behavioral Research and of the President's Commission for the Study of Ethical Problems in Medicine and Biomedical and Behavioral Research.

KAREN LEBACQZ, professor of Christian ethics at the Pacific School of Religion in Berkeley, California, is the author or editor of several books, including *Six Theories of Justice: Perspectives from Philosophical and Theological Ethics* (1986) and *The Human Embryonic Stem Cell Debate: Science, Ethics, and Public Policy* (2001). She was a member of the National Commission for the Protection of Human Subjects of Biomedical and Behavioral Research.

ROBERT J. LEVINE is professor of medicine and lecturer in pharmacology at Yale University, where he cochairs the executive committee of the Yale University Interdisciplinary Project in Bioethics. The author of *Ethics and Regulation of Clinical Research* (2nd ed., 1986), he was a special consultant to the National Commission for the Protection of Human Subjects of Biomedical and Behavioral Research.

ANNA C. MASTROIANNI is an assistant professor, University of Washington School of Law and the Institute for Public Health Genetics. She is coeditor of *Women and Health Research: Ethical and Legal Issues of Including Women in Clinical Trials* (1994) and *Beyond Consent: Seeking Justice in Research* (1998). She was associate director of the Advisory Committee on Human Radiation Experiments.

ERIC M. MESLIN is director of the Indiana University Center for Bioethics, professor of medicine and of medical and molecular genetics in the Indiana University School of Medicine, and professor of philosophy in the School of Liberal Arts. The author of a number of articles on research involving human subjects, Meslin was the executive director of the National Bioethics Advisory Commission.

HENRY S. RICHARDSON is professor of philosophy at Georgetown University and the author of *Practical Reasoning about Final Ends* (1997) and *Democratic Autonomy: Public Reasoning about the Ends of Policy* (2003) as well as coeditor of *The Philosophy of Rawls: A Collection of Essays* (1999).

HAROLD T. SHAPIRO is president emeritus and professor of economics and public affairs at Princeton University. From 1990 to 1992 he was a member of President George H. W. Bush's Council of Advisors on Science and Technology, and in 1996 he was appointed chair of the National Bioethics Advisory Commission.

SUSAN SHERWIN is fellow of the Royal Society of Canada, university research professor, Dalhousie University, Halifax, Canada. She is the editor of *The Politics of Women's Health: Exploring Agency and Autonomy* (1998) and the author of *No Longer Patient: Feminist Ethics and Health Care* (1993). She has been a consultant for the Canadian Biotechnology Advising Committee.

ROBERT M. VEATCH is professor of medical ethics, Kennedy Institute of Ethics, Georgetown University. His several books include *The Patient as Partner: A Theory of Human-Experimentation Ethics* (1987) and *The Patient-Physician Relation: The Patient as Partner, Part 2* (1991). Veatch was a consultant to the National Commission for the Protection of Human Subjects of Biomedical and Behavioral Research.

CHARLES WEIJER is Canadian Institutes of Health Investigator and associate professor of bioethics and medicine, as well as adjunct professor of philosophy, at Dalhousie University in Halifax, Canada. He has published widely on ethical issues in research involving human subjects, among other topics. He served twice as consultant to the U.S. National Bioethics Advisory Commission.

Index

Abram, Morris B., 39n.19
accounting, double-entry bookkeeping, 228–29
ACHRE. *See* Advisory Committee on Human Radiation Experiments
Ackerman, Bruce, 106
Advanced Cell Technology, 65
Advisory Committee on Human Radiation Experiments (ACHRE)
  Belmont Report, relationship to, ix–x, 41
  children, historical analysis of research involving, 44–48
  clinical research and therapeutic practices, confusion about what separates, 122
  compensatory justice, recommendations based on, 145
  contemporary research, analysis of, 48–50
  ethical framework adopted by, 42–44
  mission and context of, 41–42
  moral considerations beyond *Belmont*: trust and openness, 50–52
  multimethodological approach of, 72
  National Bioethics Advisory Commission, impact on, 55
  National Commission for the Protection of Human Subjects of Biomedical and Behavioral Research and, 41–43
  principles of, 52–53, 246
  trust, importance of, 250
AEC. *See* Atomic Energy Commission
African Americans
  inclusion in medical research, 140, 142
  life expectancy of, 157
  Tuskegee Syphilis Study. *See* Tuskegee Syphilis Study
American Medical Association (AMA), 141, 189

American Psychological Association, 189, 264n
Americans with Disabilities Act (ADA), 110n.32
Atomic Energy Commission (AEC), 46–47
Atwell, John E., 124n.14
autonomy
  Belmont Report, text of discussion in, 256–57
  clinical medicine and patient, 86–89
  conjectural history accounting for the inadequate principle of, 116–17
  consequentialist *vs.* deontological approaches to, 119–21
  continuing need for a strong principle of, 121–23
  dominance over other principles, concern regarding, 234–36
  inadequacy of *Belmont*'s interpretation of, 111–14
  proactive enablement, objections to, 114–16
  promotion of self-determination as key to robust, 112–13
  reduction to self-determination/freedom of choice, 100–101, 109n.9
  a relational approach to, 163n.9
  respect for persons and, 99–102, 128, 187–88, 247
  suggested revisions of the *Belmont Report* emphasizing, 117–19
  See also respect for persons, principle of

Baier, Kurt, 3–4, 20n.3
balancing approach to ethical principles
  appeal and problems of, 198–99
  Belmont Report, evidence of acceptance in, 195

269

CPSIA information can be obtained
at www.ICGtesting.com
Printed in the USA
LVOW05s2102080817
544257LV00021B/245/P

9 781589 010628